International Workshop Navigation and Motion Control (NMC 2021)

IOP Conference Series: Materials Science and Engineering
Volume 1215

Saint-Petersburg, Russia
20 - 24 September 2021

Editors:

Oleg Stepanov
Yulia Litvinenko

ISBN: 978-1-7138-4674-1
ISSN: 1757-8981

Printed from e-media with permission by:

Curran Associates, Inc.
57 Morehouse Lane
Red Hook, NY 12571

Some format issues inherent in the e-media version may also appear in this print version.

Institute of Physics
Dirac House, Temple Back
Bristol BS1 6BE UK

Phone: 44 1 17 929 7481
Fax: 44 1 17 920 0979

techtracking@iop.org

Additional copies of this publication are available from:

Curran Associates, Inc.
57 Morehouse Lane
Red Hook, NY 12571 USA
Phone: 845-758-0400
Fax: 845-758-2633
Email: curran@proceedings.com
Web: www.proceedings.com

TABLE OF CONTENTS

IOP Conf. Series: Materials Science and Engineering 1215 (2022) 011001 doi:10.1088/1757-899X/1215/1/011001

INTERNATIONAL WORKSHOP NMC 2021 NAVIGATION & MOTION CONTROL

20-24 September 2021, Russia

Preface

The International Workshop "Navigation and Motion Control" (NMC) is organized by the State Research Centre of the Russian Federation – Concern CSRI Elektropribor, JSC, with participation of ITMO University and St. Petersburg Electrotechnical University. Also, the Workshop would not have been held without support of the Academy of Navigation and Motion Control and its President, Academician of the Russian Academy of Sciences (RAS) Vladimir G. Peshekhonov.

The history of the Workshop originates from the annual scientific schools at CSRI Elektropribor, which were organized from 2003 to 2008 on a regular basis. Since 2010, the Workshop has been held in English, and since 2014 it has been granted the international status. Beginning from 2014, more than 200 scientists from different countries took part in the Workshop.

In 2021, the Workshop "Navigation and Motion Control" was held on 20-24 September on the picturesque shore of the Ladoga Lake.

The Workshop was attended by 54 people: students, masters, graduate students, and specialists, and experienced scientists from 5 cities of Russia and Germany. This year, due to coronavirus restrictions, participants from Germany took part in a remote format.

There were 21 presentations and 8 lectures of scientists who are authorities in their field of science. The lectures were devoted to a wide range of issues: the creation of control systems for autonomous unmanned underwater vehicles, development of autonomous ground navigation systems, emergent Intelligence and distributed stochastic optimization, and applications of the theory of elasticity in modern instrumentation. Prospects for the

IOP Conf. Series: Materials Science and Engineering 1215 (2022) 011001 doi:10.1088/1757-899X/1215/1/011001

development of navigation technologies were described in his lecture by Academician V.G . Peshekhonov.

Lectures and presentations of all the attendees were followed by lively discussions, which showed the interest of the audience and the relevance of the topics being discussed. During the Workshop, a professional translator gives presentations on how to write scientific papers.

Upon the completion of the Workshop and peer review, proceedings were prepared.

NMC-2021 covered the following topics:

- Gyroscopic systems;
- Applied problems of navigation and motion control;
- Sensors for navigation and control systems;
- MEMS sensors, systems, and technologies;
- Data processing in navigation systems;
- Integrated and satellite navigation systems;
- Control theory and control systems;
- Electronic and electromechanical devices of navigation and control systems.

International Workshop "Navigation and Motion Control" in 2021

The international **program committee** included well-known scientists from the leading institutes of Russia and Europe:

- Chairman: O. A. Stepanov, Corresponding Member of RAS, Dr. Sc., Prof. Concern CSRI Elektropribor, JSC; ITMO University, St. Petersburg, Russia;
- Co-Chairman: G. F. Trommer, Prof., PhD Institute of Systems Optimization of Karlsruhe Institute of Technology, Germany; ITMO University.

The editors

Oleg Stepanov
Yulia Litvinenko

IOP Conf. Series: Materials Science and Engineering 1215 (2022) 011002 doi:10.1088/1757-899X/1215/1/011002

Peer review declaration

All papers published in this volume of *Conference Series: Materials Science and Engineering* have been **peer reviewed through processes administered by the Editors**. Reviews were conducted by expert referees to the professional and scientific standards expected of a proceedings journal published by IOP Publishing.

- Type of peer review: Double-blind
- Number of submissions received: 27
- Number of submissions sent for review: 17
- Number of submissions accepted: 17
- Acceptance Rate (Number of Submissions Accepted / Number of Submissions Received X 100): 17/27*100 = **62,96%**
- Average number of reviews per paper: 2
- Total number of reviewers involved: 9

Any additional info on review process:

We usually use single-blind type of PR, but also in controversial cases we can use double-blind. This year we sent the paper to the reviewer for 2 weeks to make a review. Then authors made some improvements in their papers. If there was any doubt about acceptance or rejection after this step, we asked the third reviewer to analyze and make a decision. Editor rechecked the revised papers based on the gathered reviews, proofread the texts made in the Word template, edited the manuscript if needed, and checked the similarity before submitting papers in IOP. All papers meet the Technical Criteria, Quality Criteria, Presentation Criteria set by IOP Publishing.

Aims and scope

- Gyroscopic systems
- Applied problems of navigation and motion control
- Sensors for navigation and control systems
- MEMS sensors, systems, and technologies
- Data processing in navigation systems
- Integrated and satellite navigation systems
- Control theory and control systems
- Electronic and electromechanical devices of navigation and control systems

The aim of the workshop is to provide an overview of new technologies in navigation and motion control. The program includes tutorials on modern navigation and control systems as well as hardware design problems, which will be delivered both by the leading Russian and foreign scientists and young scientists.

Contact person for queries:

Dmitry O. Taranovskiy, PhD, head of editorial group
Concern CSRI Elektropribor, JSC, editor's office, St. Petersburg, Russia
editor@eprib.ru

IOP Conf. Series: Materials Science and Engineering 1215 (2022) 012001 doi:10.1088/1757-899X/1215/1/012001

Emergent Intelligence and Distributed Stochastic Optimization

O.N. Granichin[1,3], O.A. Granichina[2], V.A. Erofeeva[1], A.V. Leonova[1,3], A.A. Senov[3]

[1]IPME RAS, Saint Petersburg, Russia
[2]Herzen State Pedagogical University, Saint Petersburg, Russia
[3]Saint Petersburg State University, Saint Petersburg, Russia

o.granichin@spbu.ru

Abstract. Emergent intelligence is a property of a system of elements that is not inherent in each element individually. This behavior is based on local communications. This behavior helps to adapt to emerging uncertainties and achieve a global goal. This behavior exists in the natural world. A simplified example of emergent intelligence from the natural world is given. The repetition of natural behavior with the help of simple technical devices, which are limited in resources and cheap in construction, and the use of multi-agent approaches is considered. Distributed algorithms using local communications are considered. Such algorithms are more robust to noise.

1. Introduction

Due to the development of computer and sensor devices, interest in distributed computing and control is constantly growing. The field of distributed network systems also finds responses in practical applications, for example, in energy and social systems [1–2], mobile robotics [3]. Using multi-agent technologies has a number of advantages compared to centralized systems. The system adapts with the help of local interaction of elements in an unpredictable dynamic environment, which is not inherent in classical approaches. As well as physical restrictions on communication channels become a bottleneck of centralized systems [4]. As a result, the field of research on the properties and recreation of emergent intelligence is popular and is being studied within the framework of the interaction of mobile elements of the system [5]. The key concept in solving problems using multi-agent technologies is communication and consensus-building. This makes it possible to reach a common agreement between the elements of the system to achieve a common goal based on their own local data and data from neighboring elements. Therefore, recently, for large-scale systems involving many individual elements, distributed optimization is often used to estimate unknown parameters based on data obtained from distributed sensors [6–7]. And also especially suitable for large-scale optimization problems, a new approach is relevant, sequential subspace optimization, based on the sequential reduction of the original optimization problem to optimization problems in a low-dimensional space [8–11]. The main contribution of the aim of this work is to develop a method for distributed sequential subspace optimization.

2. Autonomous Group Behavior of Agents

Emergent intelligence (intellectual resonance, swarm intelligence) is an occurrence of unexpected properties that a system possesses, but none of its individual elements has. Key feature is the dynamics and unpredictability of the decision making process. Considering of a typical ants colony behavior helps us to illustrate scope of new approaches, multiagent technologies.

In order to avoid ambiguity and not delve into the complex processes of organizing of a living ant colony, we will consider a hypothetical colony of «artificial ants». There are two types of ants in our colony: worker ants and the main device "mother". "Mother" is able to produce new worker artificial ants. The entire study area is divided into fairly small cells. All colony needs resources (foods) to continue the living processes. Let's consider two approaches to solving the problem of finding resources.

Traditionally, there is a main element – "mother", which has super abilities, it is possible to use space satellite data to find food and to investigate an environment. After that, it will compute the optimal path and send worker ants to the food. Then all worker ants need a strong map and navigation system (e.g. using GPS). The description system is too complicated and expensive. But such a system also has one more significant drawback. If during the mission, something went wrong or something changed in the environment, then you need to promptly update the data, recalculate the plan and send new tasks to all worker ants. So, we need a fast supercomputer and a system of operational communication with worker ants.

Consider a much cheaper system without space satellite data, supercomputing for optimal path finding, maps, GPS navigation and effective communication with the main controller. This approach is multi-agent. The main element is responsible only for the production of worker ants. Each of worker ants is able to play into two missions. There are to find food and to go back to the ant hill. They have two abilities to leave pheromones in their path and to determine which of the freshest pheromones are in neighboring cells. Pheromones of two types are used. Some remain in cells when moving out of anthill without food, while others remain in cells when moving with food.

Let us illustrate the situation of the behavior of an ant colony in conditions of uncertainty (Fig. 1).

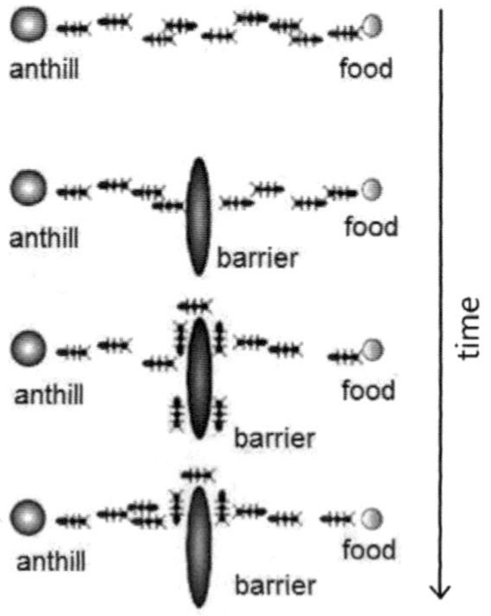

Figure 1. The ant colony behavior under conditions of uncertainty. Optimization of the path

Let the road be from anthill to food. If it is broken after a storm (some catastrophic event) then ants' colony needs a new one. The strategy is to spread ants in all possible directions. For example, they can be chosen randomly. It is a randomized strategy. This strategy allows to find new ways to the food and to optimize the path. This system is cheaper than the previous one and it remains workable and semi-optimal under unpredictable uncertainties. The new term "Emergent Intelligence" is used for these kinds of behaviors.

The paper [12] presents a new interpretation of emergent intelligence as a problem of multi-agent technologies. Using examples of practical NP-hard problems, a parallel is drawn between multi-agent technologies and emerging intelligence.

A global problem and a system where there are many simple devices are considered. The article provides a detailed analysis of the impact of local communication on achieving a global goal on the example of a swarm of cheap devices that do not have the resources to build a global map of the world. The analysis is an optimization of the parameter α from the formula of the robot's confidence in its actions. With the parameter value equal to zero, the system's devices do not interact locally, and with the parameter value equal to one, the devices mainly rely not on their own measurements but on noisy data of neighbors. The analysis is an optimization of the parameter α from the formula of the robot's confidence in its actions. With the parameter value equal to zero, the system's devices do not interact locally, and with the parameter value equal to one, the devices mainly rely on noisy data from neighbors, and not on their own measurements. The results of the analysis show that the optimal value of the parameter is approximately 0.8. This shows that solving a global problem is more robust and faster when using local communication than without interaction between devices.

Thus, with this approach, the global problem is solved even under the condition of significant disturbances. As a result, this behavior can be called emergent intelligence.

3. Distributed Stochastic Approximation under Unknown-But-Bounded Disturbances

A similar idea of solving the practical problem of multi-agent technologies with unknown-but-bounded perturbations is discussed in the article [13]. A new distributed algorithm based on stochastic approximation of simultaneous disturbances (SPSA) is proposed. The goal of distributed optimization is (usually) to compute some cost function minimizer by utilizing an interaction among the agents

$$\bar{F}(x) = \sum_{i=1}^{n} F^i(x),$$

where $x \in \mathbb{R}^d$, $F^i(x): \mathbb{R}^d \to \mathbb{R}$ is the cost function of agent i, known only to it.

Let's define the topology of a multi-agent system. Interaction within a system consisting n elements can be described using a directed graph $G = (N, E)$, where $N = \{1, \dots, n\}$ is a set of vertices, $E \subseteq N \times N$ is the set of directed edges. Sub-graph $\overline{G}$ is a graph $\overline{G} = (N_{\overline{G}}, E_{\overline{G}})$, where $N_{\overline{G}} \subseteq N$ and $E_{\overline{G}} \subseteq E$. For node $i \in N$, the set of neighbors is $N^i = \{j \in N : (j,i) \in E\}$. Node $i \in N$ in-degree is $|N^i|$. For each arc $(j,i) \in E$ we set weight $c^{i,j} > 0$ and $c^{i,j} = 0$ for $(j,i) \notin E$. Denote $C = [c^{i,j}]$ weighted adjacency matrix or connectivity matrix. $G_C = (N_C, E_C)$ is a weighted digraph, where $N_C \equiv N$ and $E_C = E$. Let $c^{i,j}$ be the cost of data transfer through directed edge $(j,i) \in E_C$. Weighted in-degree of node $i \in N_C$ equals $\deg_i^+(j,i) \in E_C$, $\deg_{\max}^+(C)$ is the maximal in-degree among all nodes in graph G_C.

Cost constraints for node $i \in N_C$ can be specified as an upper bound Q: $\deg_i^+(C) \le Q$, which can denote the total "cost" of agent i communication with its neighbors. To satisfy the constraints, it is possible to generate at time t the sub-graph $G_{B_t} \subset G_C$, specified by the weighted adjacency matrix B_t so that $\deg_i^+(B_t) \le Q$. Cost constraints $\deg_i^+(B_t) \le Q$ may not always be satisfied for given $B_t = C$ and Q, for example, for $n = 6$, complete graph G_C, $c^{i,j} = 1$, $i \ne j$, $c^{i,i}$ and Q < 5. One possible solution is to use a randomized topology where $5 - Q$ directed edges are randomly excluded. With $Q = 1$ this strategy is similar to the one used in algorithms like gossip (gossip algorithms) [14]. Let's give a definition: A random sub-graph G_{B_t} satisfies the cost constraints with the level Q if

$$\mathbb{E} \deg_{\max}^+(B_t) \le Q \qquad (0.1)$$

Denote by Ξ the set, $\{f_\xi^i(\theta)\}_{\xi\in\Xi}$ family of differentiable functions: $\forall i \in N$ $f_\xi^i(\theta):\mathbb{R}^d \to \mathbb{R}$. It is assumed that the parameter θ cannot be measured directly. Therefore, a sequence of measurement points $x_1^i, x_2^i, \ldots, i \in N$ is selected according to the observation plan. The values $y_1^i, y_2^i, \ldots$ of functions $f_\xi^i(\cdot)$ are obtained at time instants $t = 1,2,\ldots$ with additive external unknown but bounded noise v_t^i

$$y_t^i = f_{\xi_t}^i(x_t^i) + v_t^i, \tag{0.2}$$

where $\{\xi_t\}, \xi_t \in \Xi$ is an uncontrolled (e.g., $\Xi = \mathbb{N}$ and $\xi_t = t$) or a random sequence. In the latter case, it is assumed that there is a probability distribution ξ_t, which may be known or unknown. Let $\mathcal{F}_{t-1}$ be the σ-algebra of all random events that occurred before the time *t*, $\mathbb{E}_{\mathcal{F}_{t-1}}$ denote the conditional expectation with respect to σ-algebra $\mathcal{F}_{t-1}$. Consider an optimization problem in which the cost function $\overline{F}_t(\theta)$ is expressed as a sum of local functions $\overline{F}_t(\theta) = \mathbb{E}_{\mathcal{F}_{t-1}} f_{\xi_t}^i(\theta)$ each of which depends on the common optimization variable θ. In addition, the minimum point θ of the function $\overline{F}_t(\theta)$ may change over time. Optimization of the average risk functional: estimate the time-varying minimum point θ_t of the distributed function

$$\overline{F}_t(\theta) = \sum_{i\in N} F_t^i(\theta) = \mathbb{E}_{\mathcal{F}_{t-1}} \sum_{i\in N} f_{\xi_t}^i(\theta) \to \min_\theta.$$

Suppose at time instant *t* the sensors can communicate with their neighbors over the network defined by the graph $G_{B_t} = (N_{B_t}, E_{B_t})$. The corresponding adjacency matrix B_t satisfies some averaged cost constraints (0.1) with the level Q. Assume also that sensor *i* receives the current measurement $\widehat{\theta}_t^i$ with noise (0.2) and, if the set $N_t^i = \{j \in N_{B_t} : (j,i) \in E_{B_t}\}$ is not empty, it also receives current estimates transmitted by neighbors through noisy communication channels $\tilde{\theta}_t^{i,j} = \hat{\theta}_t^j + w_t^{i,j}, j \in N_t^i$, where $w_t^{i,j}$ is a noise in the communication channel. If $j \notin N_t^i$ then $\tilde{\theta}_t^{i,j} = 0$.

Introduce $\Delta_k^i, k = 1,2,\ldots, i \in N$ sequence of independent random vectors from $\mathbb{R}^d$, called simultaneous trial perturbation, with symmetric distribution functions $P_k^i(\cdot)$; and set of vector-functions (kernels) $K_k^i(\cdot):\mathbb{R}^d \to \mathbb{R}^d, k = 1,2,\ldots$. Choose initial approximation $\hat{\theta}_0^i \in \mathbb{R}^d$, positive step α, gain γ, and sequences of non-negative numbers $\{\beta_k^+\}$ and $\{\beta_k^-\}$ such that $\beta_k = \beta_k^+ + \beta_k^- > 0$. Algorithm with two measurements of functions $f_{\xi_t}^i(\theta)$ for each agent $i \in N$, constructing a sequence of measurement points $\{x_t^i\}$ and estimates $\{\hat{\theta}_t^i\}$:

$$\begin{cases} x_{2k-1}^i = \hat{\theta}_{2k-2}^i - \beta_k^- \Delta_k^i, \\ x_{2k}^i = \hat{\theta}_{2k-2}^i + \beta_k^+ \Delta_k^i, \\ \hat{\theta}_{2k-1}^i = \hat{\theta}_{2k-2}^i, \\ \hat{\theta}_{2k}^i = \hat{\theta}_{2k-1}^i - \alpha\left(\dfrac{y_{2k}^i - y_{2k-1}^i}{\beta_k} K_k^i(\Delta_k^i) + \gamma \sum_{j\in N_{2k-1}^i} b_{2k-1}^{i,j} (\tilde{\theta}_{2k-1}^{i,j} - \hat{\theta}_{2k-1}^i) \right) \end{cases} \tag{0.3}$$

4. Distributed Subspace Optimization

4.1. Algorithm's description

The article [11] proposes an algorithm for sequential subspace optimization for a convex large-scale optimization problem. The main idea of the method is to reduce the problem of the original space to a problem in some subspace of much smaller dimension. This algorithm proved to be better than some other popular algorithms: Broyden-Fletcher-Goldfarb-Shanno with limited memory, conjugate gradient and Beale- Powell conjugate gradients methods. For problems with significant interference, a distributed algorithm for sequential subspace optimization is proposed.

Let $G_A = \{N, E\}$ be a graph representing the network topology associated with adjacency matrix A, where $N = \{1,...,n\}$ is a set of nodes, and $E \subseteq N \times N$ is a set of edges. Denote by $i \in N$ an identifier of i-th node and $(j,i) \in E$ if there is a directed edge from node j to node i. The latter means that node j is able to transmit data to node i. For a node $i \in N$, the set of neighbors is defined as $N^i = \{j \in N : (j,i) \in E\}$.

We consider the following optimization problem:

$$\min_{\bar{x} \in \mathbb{R}^d} \left[F(\bar{x}) = \sum_{i=1}^{n} f^i(x^i) \right]. \tag{0.4}$$

Decision vector $\bar{x}$ consists of n sub-blocks, i.e., $\bar{x} = [x^1,...,x^n]^T$. Node i optimizes only sub-block x^i. Each sub-block include two types of variables. The first one is observable exclusively by node i, and its variables belong to set $\mathcal{X}_i$. The second one is shared with the nodes that observe the same variables, and its variables belong to sets $\mathcal{S}_{i,j}, i \in N, j \in N^i$.

Since functions depend only on sub-blocks of variables, we don't need to exchange full vectors between them. Consider n sub-networks, which we assume to be connected and undirected. For each network we define symmetric Laplacian matrix $W_{i,j}$, and vector of variables that is common for both nodes i and j, i.e., $x^{i,j}$. Based on these notations we reformulate problem (0.4) to achieve a common solution $\bar{x}^*$ through consensus as follows:

$$\min_{\bar{x} \in \mathbb{R}^d} \left[F(\bar{x}) = \sum_{i=1}^{n} f^i(x^i) \right], \tag{0.5}$$

$$\text{subject to } W_{i,j} x^{i,j} = 0, \forall i \in N, j \in N^i \tag{0.6}$$

Assume that the gradients ∇f^i are known. Using the conventional L-QNSSO method published in [11] we propose its distributed version. At each node i produce the following sequence of steps:

1) $D_t^i \leftarrow subspaseUpdate(D_{t-1}^i, \nabla f^i(x_t^i), m^i)$
2) if $t = 0$ then $d_t^i \leftarrow -\nabla f^i(x^i)$ and go to step 10
3) $T^i \leftarrow \min(t, m^i - 1)$
4) $Z_t^i \leftarrow [[D_t^i]^T (x_t^i - x_{t-1}^i),...,[D_t^i]^T (x_t^i - x_{t-T}^i)]^T$
5) $G_t^i \leftarrow [[D_t^i]^T (\nabla f^i(x_t^i) - \nabla f^i(x_{t-1}^i)),...,[D_t^i]^T (\nabla f^i(x_t^i) - \nabla f^i(x_{t-T}^i))]^T$
6) $d_t^{i,QN} \leftarrow -D_t^i([G_t^i]^T G_t^i)^{-1}[G_t^i]^T Z_t^i [D_t^i]^T \nabla f^i(x_t^i)$
7) $\beta_t^i \leftarrow \dfrac{\nabla f^i(x_t^i)^T (\nabla f^i(x_t^i) - \nabla f^i(x_{t-1}^i))}{\nabla f^i(x_{t-1}^i)^T \nabla f^i(x_{t-1}^i)}$
8) $d_t^{i,CG} \leftarrow (I - D_t^i[G_t^i]^T (G_t^i[G_t^i]^T)^{-1} G_t^i [D_t^i]^T)$

9) $d_t^i \leftarrow d_t^{i,QN} + d_t^{i,CG}$

10) $\alpha_t^i \leftarrow \arg\min f^i(x_t^i + \alpha^i d_t^i)$

11) Collect x_t^j that that belongs to $\mathcal{S}_{i,j}$ from neighbors

12) $x_{t+1}^i \leftarrow x_t^i + \alpha_t^i d_t^i$

13) $\forall j \in N^i, x_{t+1}^{i,j} \leftarrow x_t^{i,j} - \gamma^i (x_t^{i,j} - x_t^j)$

14) $t \leftarrow t+1$; go to step 1

4.2. *Simulation*

Numerical experiments are presented that illustrate the performance of the proposed algorithm DL-QNSSO. The algorithm is applied to the BAL dataset [15]. The algorithm is tested at different noises. There are gaussian distribution (disturb_normal), pareto distribution (disturb_pareto), uniform distribution (disturb_uniform), Gaussian distribution with an non-diagonal covariance matrix and a large eccentricity (disturb_multinormal) and mixture of gaussians (disturb_gauss_mixture).

The BAL Datasets, e.g. Trafalgar, contain several problems. For each problem we generate 5 noisy datasets and make 10 different runs. The different runs mean that we fix a distribution and its parameters, and generate 10 noise sequences with different seeds. In total, we have 50 results per problem. Figure 2 demonstrate the general result of experiments with noise analysis.

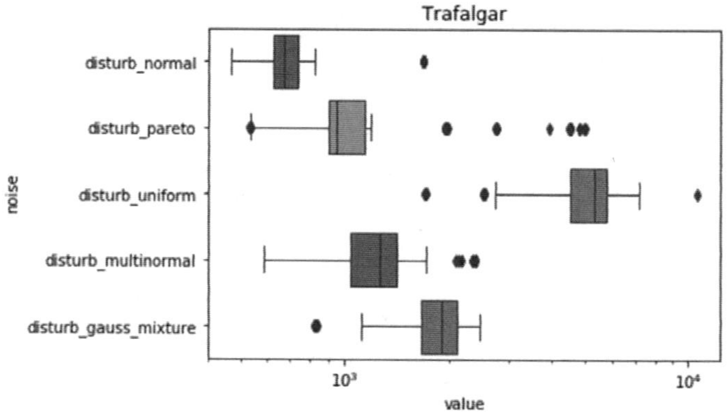

Figure 2. Trafalgar dataset. Numerical experiments results with different noises

This graphic is built for Trafalgar dataset. Each boxplot accumulates values for each noise for the problems contained in a dataset. The black dots on the graph are outliers.

5. Conclusions

The illustrated example shows the concept of emergent intelligence. The connection between emergent intelligence and multi-agent technologies is described. Algorithms of distributed computing and control for the problem of estimating unknown parameters for a multi-agent system are considered. A new distributed sequential subspace optimization algorithm suitable for large-scale optimization is proposed. Numerical experiments for the proposed algorithm on an open dataset are given.

References

[1] Ren, W., Beard, R.W. Distributed consensus in multi-vehicle cooperative control, London: Springer London, 2008, vol. 27, issue 2, pp. 71–82.

[2] Lewis, F.L., Zhang, H., Hengster-Movric, K., Das, A. Cooperative control of multi-agent systems: optimal and adaptive design approaches, Springer Science & Business Media, 2013.

[3] Savla K., Notarstefano G., Bullo F. Maintaining limited-range connectivity among second-order agents, SIAM Journal on Control and Optimization, 2009, vol. 48, issue 1, pp. 187–205.
[4] Grachev, S., Skobelev, P., Mayorov, I., Simonova, E. Adaptive clustering through multi-agent technology: Development and perspectives, Mathematics, 2020, vol. 8, issue 10, p. 1664.
[5] Vicsek, T., Zafeiris, A. Collective motion, Physics reports, 2012, vol. 517, issue 3-4, pp. 71–140.
[6] Aeron, S., Saligrama, V., Castanon, D.A. Efficient sensor management policies for distributed target tracking in multihop sensor networks, IEEE Transactions on Signal Processing, 2008, vol. 56, issue 6, pp. 2562–2574.
[7] Sayed, A.H., Tu, S.-Y., Chen, J., Zhao, X., Towfic, Z.J. Diffusion strategies for adaptation and learning over networks: an examination of distributed strategies and network behavior, IEEE Signal Processing Magazine, 2013, vol. 30, issue 3, pp. 155–171.
[8] Conn, A.R., Toint, P.L., Sartenaer, A., Gould, N.I.M. On iterated-subspace minimization methods for nonlinear optimization, Linear and Nonlinear Conjugate-Gradient Related Methods, SIAM, 1994, pp. 50–78.
[9] Yuan, Y.X. Subspace methods for large scale nonlinear equations and nonlinear least squares, Optimization and Engineering, 2009, vol. 10, issue 2, pp. 207–218.
[10] Chouzenoux, E., Idier, J., Moussaoui, S. A majorize–minimize strategy for subspace optimization applied to image restoration, IEEE Transactions on Image Processing, 2010, vol. 20, issue 6, pp. 1517–1528.
[11] Senov, A.A., Granichin, O.N., Granichina, O.A. A Sequential Subspace Quasi-Newton Method for Large-Scale Convex Optimization, 2020 American Control Conference (ACC), IEEE, 2020. pp. 3627–3632.
[12] Amelin, K., Granichin, O.N., Sergeenko, A., Volkovich, Zeev V.. Emergent Intelligence via Self-Organization in a Group of Robotic Devices, Mathematics, 2021, vol. 9, issue 12, p. 1314.
[13] Granichin, O.N., Erofeeva, V.A., Ivanskiy, Y.V., Jiang, Y. Simultaneous Perturbation Stochastic Approximation-based Consensus for Tracking under Unknown-but-Bounded Disturbances, IEEE Transactions on Automatic Control, 2021, vol. 66, issue 8, pp. 3710–3717.
[14] Boyd, S., Ghosh, A., Prabhakar, B., Shag, D. Randomized gossip algorithms, IEEE transactions on information theory, 2006, vol. 52, issue 6, pp. 2508–2530.
[15] Site: https://grail.cs.washington.edu/projects/bal/ (online; accessed: 04.08.2021)

Acknowledgements

The research was carried out within the IPME RAS grant supported by Russian Science Foundation (project No. 21-19-00516) (Sections Autonomous Group Behavior of Agents, Distributed Subspace Optimization) and the SPbSU grant supported by Russian Foundation for Basic Research (project No. 20-01-00619) (Section Distributed Stochastic Approximation under Unknown-But-Bounded).

IOP Conf. Series: Materials Science and Engineering 1215 (2022) 012002 doi:10.1088/1757-899X/1215/1/012002

Comparative Analysis of Pitching Prediction Algorithms

D. Antonov, O. Zaitsev, Yu. Litvinenko

Concern CSRI Elektropribor, JSC, Saint Petersburg
ITMO University, Saint Petersburg

danila.v.antonov@mail.ru

Abstract. Two algorithms are described in the paper; one of them is the Kalman filter, which is based on the use of a pitching mathematical model, and the second uses a neural network in which the model is considered unknown. The results of the algorithms sensitivity analysis to the parameters of the model and its influence on the potential accuracy of prediction are presented. A stationary narrow-band second-order Markov process is used as a model of the ship pitching, which was used to form the input signal of the algorithms. Also, the results of the algorithms simulation in predicting real data are presented.

1. Introduction

It is often required to carry out work, such as stabilizing the antenna equipment, landing on deck, goods transfer, on the offshore facilities in high seas. To increase the accuracy when carrying out such works, it is required to provide a short-term pitch prediction.

All pitch prediction algorithms can be divided into two types: with and without using a pitching mathematical model. A traditional example for solving a prediction problem using a model is the Kalman filter (KF) algorithm [1-3], synthesized under the assumption that pitching is described as a random process. As algorithms in which the pitching model is not directly used, one can single out algorithms based on machine learning methods, in particular, such as decision trees and neural networks (NN) [4-9]. Working with the algorithms included in the second group is divided into training, testing and the prediction mode. Even though training usually requires a lot of computing power and time resources, it can be done in advance in off-line work. In this case, an algorithm that is simple to implement usually works in the on-line mode. In this paper, we study the sensitivity of two types of algorithms to the parameters of the real process. As a measure of the algorithm efficiency, a root mean square error is used.

2. Pitch prediction problem statement

It is assumed that the ship oscillations (pitch or roll) $u(t)$ are described by a zero-mean Gaussian stationary random process with a correlation function of the form [10]:

$$K_u(\tau) = \sigma^2 e^{-\alpha|\tau|}\left(\cos(\beta\tau) + \frac{\alpha}{\beta}\sin(\beta|\tau|)\right) \tag{1}$$

Scientific adviser: D.A. Koshaev, Ph.D (Concern CSRI Elektropribor, JSC, Saint Petersburg)

where α, β – coefficients defining the temporal properties of the process, σ – standard deviation of the process. Following [11], these coefficients for pitching will be determined using the equation below:

$$\beta = 0.98\omega_0, \ \alpha = 0.21\beta, \tag{2}$$

where ω_0 – base process frequency, determined as $\omega_0 = \sqrt{\alpha^2 + \beta^2}$.

It is known that such a process is a narrow-band second-order Markov process and its sample can be obtained using a shaping filter

$$\dot{x}(t) = Fx(t) + Gw(t), \tag{3}$$

where $x(t)$ – a two-dimensional vector whose first component is the pitch, i.e. $u(t) = Hx(t)$, where $H = [1 \ \ 0]$; $w(t)$ — Gaussian zero-mean white noise whose power spectral density (PSD) is unity,

$$F = \begin{bmatrix} 0 & 1 \\ -(\alpha^2 + \beta^2) & -2\alpha \end{bmatrix}, \ G = \begin{bmatrix} 0 & 2\sigma\sqrt{\alpha(\alpha^2 + \beta^2)} \end{bmatrix}^T.$$

Initial values $x(0)$ are a zero-mean Gaussian random vector with a covariance matrix

$$P(0) = \begin{bmatrix} \sigma^2 & 0 \\ 0 & \sigma^2(\alpha^2 + \beta^2) \end{bmatrix}. \tag{4}$$

Let us assume that there are measurements:

$$y(t) = Hx(t) + v(t) = u(t) + v(t), \tag{5}$$

where $v(t)$ – zero-mean white Gaussian measurement noise, with PSD equal to R, independent of $w(t)$ and $x(0)$.

Having measurements on the $[t - \tau_{mes}; t]$ interval, it is required to obtain the optimal, in the mean-square sense, estimate of the pitch prediction $\hat{u}(t + \tau_{pr})$, where τ_{mes} – measuring interval, τ_{pr} – prediction interval.

An important feature when solving the prediction problem is the fact that the pitching can be described by a stationary process. This means that the parameters σ and ω_0, determining the standard deviation of the wave ordinate and base frequency, do not change during the time of data processing and prediction [11].

3. Prediction algorithms

An algorithm using a KF-based pitching model and an algorithm in the form of a simple non-recurrent NN were chosen as prediction algorithms for comparison [12].

The estimate of the KF-based prediction for a pitching model in the form of a narrow-band second-order Markov process is determined using the following equation:

$$\hat{u}(t + \tau_{pr}) = H\phi(\tau_{pr})\hat{x}^{KF}(t), \tag{6}$$

where $\hat{x}^{KF}(t)$ – KF estimation for the state vector, described by the equation (3) by measurements (5), $\phi(\tau_{pr})$ – fundamental matrix of the shaping filter, defined as

$$\phi(\tau_{pr}) = e^{-\alpha|\tau_{pr}|}\begin{bmatrix} \cos\beta\tau_{pr} + \frac{\alpha}{\beta}\sin\beta|\tau_{pr}| & \frac{1}{\beta}\sin\beta\tau_{pr} \\ -\frac{(\alpha^2+\beta^2)}{\beta}\sin\beta\tau_{pr} & \cos\beta\tau_{pr} - \frac{\alpha}{\beta}\sin\beta|\tau_{pr}| \end{bmatrix}. \quad (7)$$

When obtaining an estimate in the form (6), measurements (5) accumulated over the time interval τ_{mes} preceding the prediction time t, i.e. the time interval equal to $(t-\tau_{mes};\ t)$, are used as input data for the Kalman filter. The problem is solved in discrete time using standard sampling procedures with a sampling time dt [10]. Thus, the total number of measurements when solving the problem in discrete time is $L = \frac{\tau_{mes}}{dt}$. When carrying out calculations, the sampling time was taken equal to 0.01sec.

As a prediction algorithm, in which the model is not used directly, as noted above, the simplest non-recurrent neural network was chosen (Fig.1).

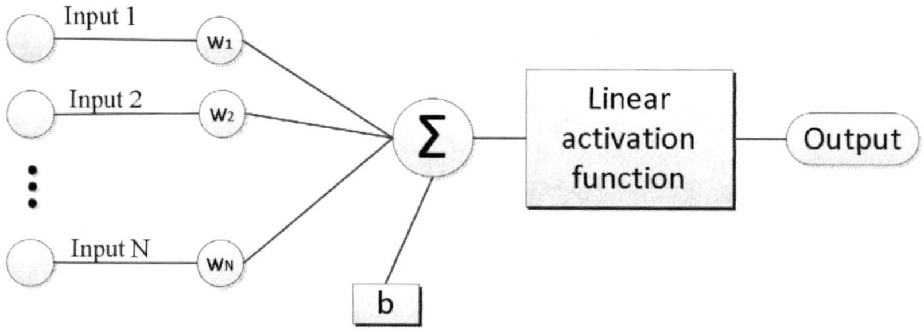

Figure 1. The architecture of the studied NN, where N=20 – number of inputs, $w_{1..N}$ – weight coefficients, b – bias.

The chosen neural network has an architecture with 20 inputs, one internal neuron and 1 output. To ensure the operation of such a NN, it is necessary to train it, i.e. find weights w_N (N = 20) and bias b. Together with the learning process of the NN, it is configured, i.e. hyperparameters are experimentally found (number of epochs, type of learning algorithm and learning rate) [13].

Let us explain how the NN is trained. We assume that a training sample is given on the time interval T_{tr} =1000 *sec*. In order to train the neural network using the training sample, sets of input data (samples) are formed (Fig. 2), including the values of the training sample on the interval $(t-\tau_{mes},\ t)$. Moreover, for each such set a value is formed corresponding to the time $t+\tau_{pr}$, i.e. for τ_{pr} ahead with respect to the moment t. It is important to emphasize that this value is used as a reference value with which the output of the NN is compared during training. It follows that it is highly desirable that the reference value used should not contain measurement errors.

With each addressing to the NN during its training, the values of the pitch are received at the input, accumulated within 1 second, which, with sampling time dt = 0.05 sec corresponds to 20 pitch values: the current one and 19 previous. Thus, the NN takes into account all the pitch values for the last second. At the output of the network, the predicted pitch value should be formed $\hat{u}(t+\tau_{pr})$ for τ_{pr} *sec* ahead.

During training, 19980 samples were sequentially received at the input of the network, and at the same time the same number of predicted values and their errors were formed at the output of the NN in the form of differences between the obtained and reference values. After processing the last sample, the weights were corrected by the backpropagation method in order to minimize the quadratic loss function. This procedure is called the learning epoch [12, 13]. Then the whole procedure was repeated using the same training sample. In addition, the learning rate was adjusted, i.e. determination of the coefficient by which the gradient of the loss function is multiplied for the corresponding weights. This is done for more accurate finding of the local minimum of the loss function [12]. In order to avoid the appearance of a high-frequency noise component in the prediction, each input value was normalized by subtracting the average value calculated over the interval τ_{mes} and division by a standard deviation of the training sample calculated on the same interval. The initial values of the weights and bias were formed as random variables using a random number generator uniformly distributed within [-0.1; 0.1]. Thus, it is necessary to get 400 different NN to ensure the possibility of predicting on an interval of 20 sec with a step $dt = 0.05$ sec.

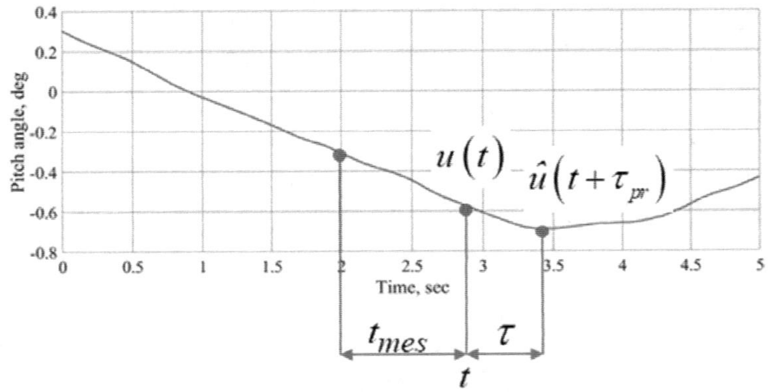

Figure 2. An example of implementing a fragment of a training sample used to form a set of input data

4. Simulation

During research the training sample was generated using a shaping filter (3) with duration T_{tr}, a dominant period $T_0 = 2\pi/\omega_0 = 10$ sec, discreteness dt=0.05 sec and σ=0.5 deg. The simulation results were obtained using the Matlab software package, including [14].

All graphs below show the root mean square error (RMSE) of the pitch prediction, averaged over 1000 samples, for prediction algorithms based on KF and NN. In Fig. 6 – 9 on the abscissa axis the ratio of the prediction time to the pitch period, % is depicted.

Fig. 3 shows the RMSE of the prediction for the KF without measurement noise for a different number of used measurements L.

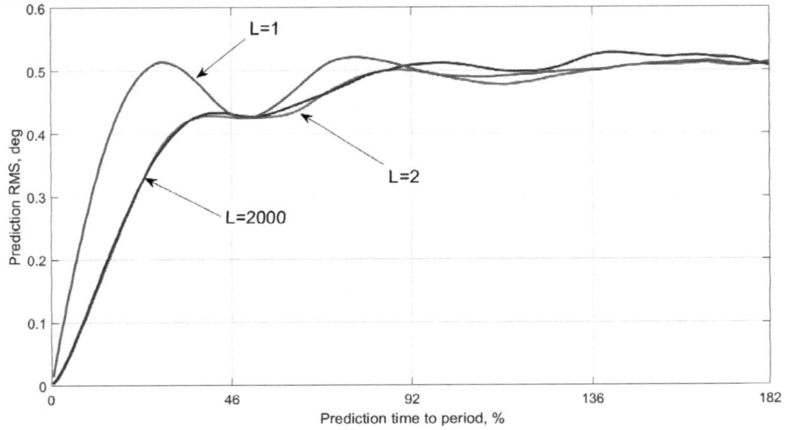

Figure 3. RMSE of prediction using KF in the absence of measurement errors at L=1, L=2, L=2000. Y-axis - prediction RMSE, deg.

PSD of measurement noise was taken equal to $R = 10^{-10}$ deg^{2}sec, which, in discrete time realization, corresponded to the standard deviation ($\sigma_n = \sqrt{R/dt}$ deg). In this case, the KF was tuned to the base frequency $\omega_0 = 0{,}6$ rad/sec ($\omega_0 = \sqrt{\alpha^2 + \beta^2}$) and σ = 0,5 deg. From Fig. 3 it follows that without measurement noise with L=1 the prediction accuracy is lower than with L=2. Results for L=2 and L=2000 practically do not differ, which is explained by the properties of a second-order Markov process, for which the values in the future are determined by the most recent values in the past. It should be noted that even without any measurement noise after the prediction time close to the value of the pitch period, the RMSE reaches the a priori level $\sigma = 0{,}5$ deg, which indicates the inexpediency of any prediction at such a time for the considered random process.

Fig. 4 (a) shows a set of plots for the RMSE obtained using the KF for different values of the measurement noise PSD equal to R. With measurement noise as $\sigma_n = \sqrt{R/dt} = 0.45$ deg ($R = 10^{-2}$ deg^{2}sec) (curve 2), which is comparable to the standard deviation of the simulated pitching, to obtain a prediction using the KF with maximum accuracy, L=50 measurements are required ($\tau_{mes} = 2.5$ sec). Otherwise, from the values $\sigma_n \geq 0.45$ deg, RMSE at the first step of the prediction increases and becomes several times higher than at a low noise level (curve 3). It should be noticed that in case of measurements with noise, the time, for which a prediction is advisable to be made, is reduced.

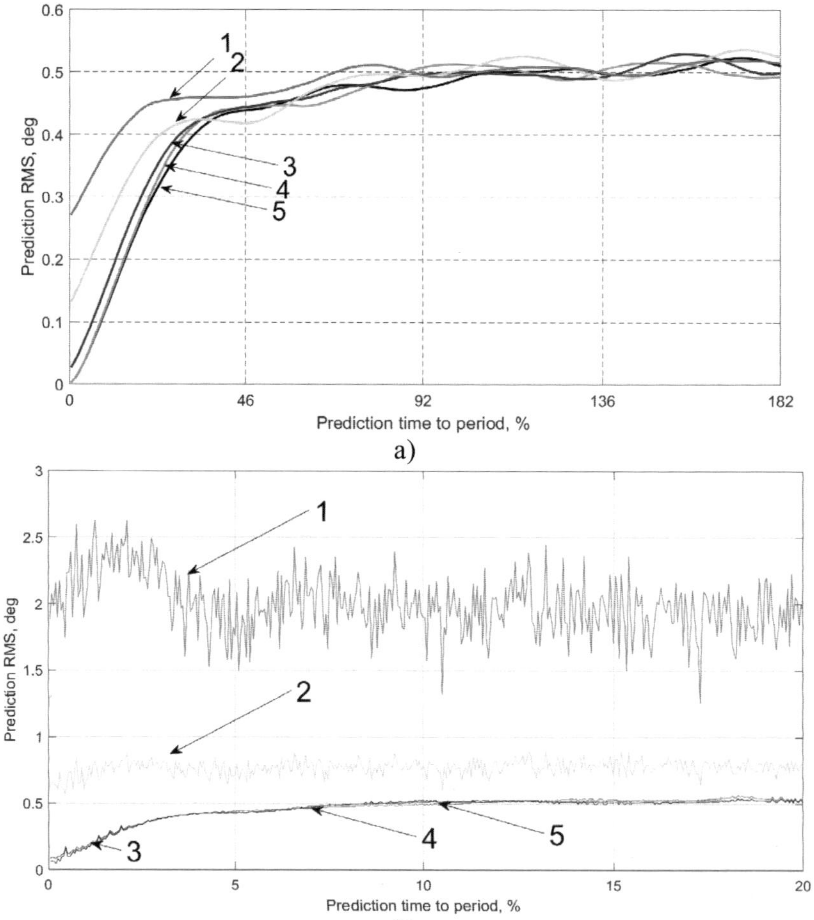

b)

Figure 4. RMSE of the prediction for the samples with measurement errors at different values of the RMS of the measurement noise (deg): a) KF, b) NN. $\sigma_n = 1.41$ deg – 1, $\sigma_n = 4.5 \cdot 10^{-1}$ deg – 2, $\sigma_n = 4.5 \cdot 10^{-2}$ deg – 3, $\sigma_n = 4.5 \cdot 10^{-3}$ deg – 4, $\sigma_n = 4.5 \cdot 10^{-5}$ deg – 5

The results for different values of the RMSE of discrete noise for the NN are presented in Fig. 4 (b). The figure shows that with the standard deviation of the noise $\sigma_n = 0.45\,\text{deg}$ (curve 2), comparable to the standard deviation of the simulated pitching, the NN-based algorithm does not provide a prediction, and the error becomes much larger than the standard deviation of the process - at the level of two degrees.

To study the sensitivity, i.e. dependence of the results of both algorithms on deviations of the actual base frequency from the calculated value $\omega_0 = 0.6$ rad/sec, four identical sets of training sample were used with duration T_{tr} and differing in the actual value of the base pitch frequency: $\omega_1 = 0.1$ rad/sec, $\omega_2 = 0.3$ rad/sec, $\omega_3 = 1.2$ rad/sec, $\omega_4 = 2$ rad/sec. In this case, the algorithms were tuned to the calculated value ω_0 (NN training was carried out on a training sample corresponding to the frequency $\omega_0 = 0.6$ rad/c). The results are shown in Fig. 5.

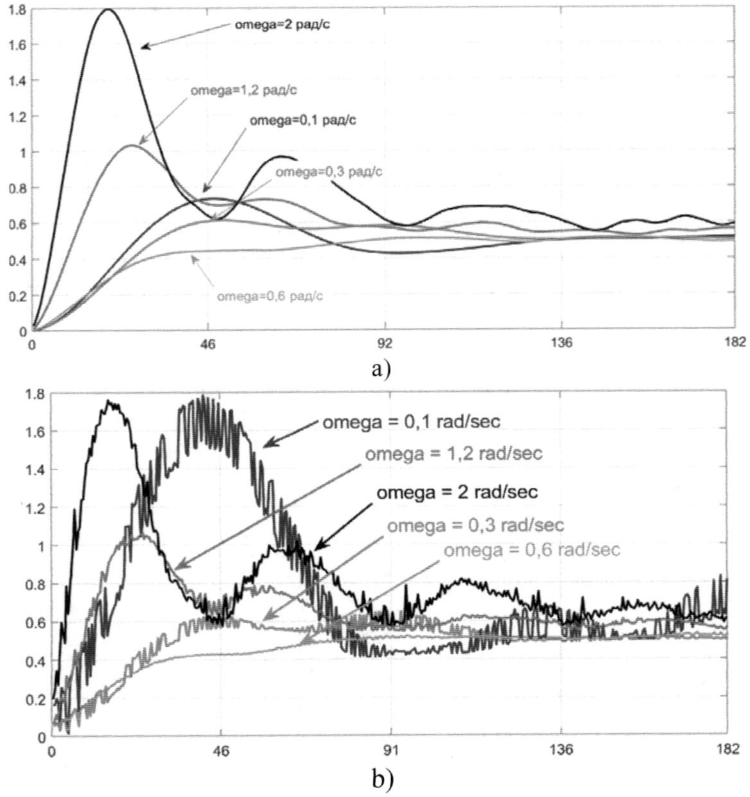

Figure 5. RMSE of the pitch prediction for various values of the base frequency (rad/s) at its calculated value equal to $\omega_0 = 0{,}6$ rad/sec: a) KF; b) NN. $\omega_0 = 0.6$ rad/sec, $\omega_1 = 0.1$ rad/sec, $\omega_2 = 0.3$ rad/sec, $\omega_3 = 1.2$ rad/sec, $\omega_4 = 2$ rad/sec.

From Fig. 5 (a), it follows that, as expected, the maximum value of the KF prediction RMSE for the frequency, to which the filter is tuned, is the smallest one. It should also be noted that with overestimated values of the actual frequency for the training sample relative to the one, to which the KF was tuned (calculated frequency), the prediction RMSE reaches the maximum in the first seconds of the prediction ($\omega_3 = 1.2$ rad/sec, $\omega_4 = 2$ rad/sec in Fig. 5(a)). The loss in the prediction accuracy with a sixfold decrease and a threefold increase of the actual frequency of the training sample relative to the calculated one is 1.5 - 3 times.

From the results shown in Fig. 5 (b), it can be seen that the NN is also sensitive to changes in the wave parameters, especially when the actual frequency decreases relative to the calculated value ω_0: accuracy decreases even more than in the case of using KF. When the frequency increases to 2 rad/s, the accuracy decreases by 3 times.

5. Simulation with real data

The data obtained from the ship's inertial navigation system were used as real data. The standard deviation of pitch was equal to $\sigma = 0.2$ deg, and the PSD value of the measurement noise was taken to be close to zero $R = 10^{-9}$ deg^{2}sec. Fig. 6 (a) presents the prediction RMSE for the KF: real RMSE, obtained from 1000 samples and calculated RMSE $\sigma_{teor}(t_k)$, determined by the formulas

$$S(t_k) = F_d S(t_k - 1) F_d^T + G_d Q_d G_d^T, \tag{8}$$

$$\sigma_{teor}(t_k) = \sqrt{S_{1,1}(t_k)}, k = 1,2 \dots \tau_{pr} / dt, \tag{9}$$

where t_k - current discrete moment in time, $S_{1,1}(t_k)$ - estimation errors covariance matrix first element, $S_0 = P^*$, P^* corresponds to the covariance matrix of the estimation errors in the KF at the moment t_0 when the last measurement is received before the start of the prediction, $F_d \approx E + F \cdot dt = \begin{bmatrix} 1 & dt \\ -(\alpha^2 + \beta^2)dt & 1 - 2\alpha dt \end{bmatrix}$, $G_d \approx G dt$, $Q_d = 1/dt$.

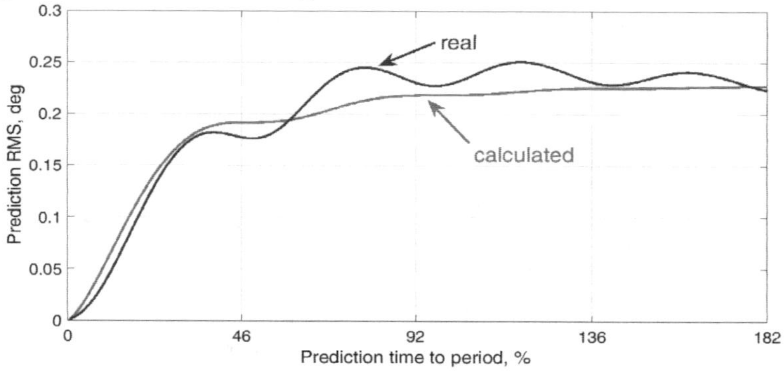

a)

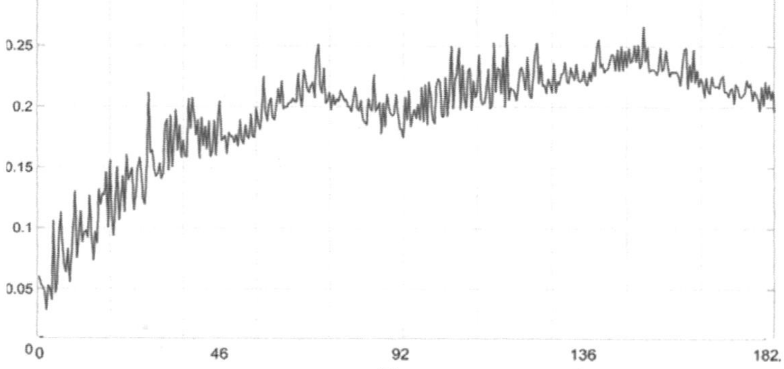

b)

Figure 6. a) Prediction RMSE using KF based on real data: real, calculated
b) prediction RMSE using NN based on real data

When calculating (8), the transition to the discrete form of the KF was carried out using the standard formulas [10].

Fig. 6 (b) shows the selective RMSE of the prediction for the NN obtained from 1000 samples. In this case, the prediction algorithm is trained on samples formed from real data according to the previously described method. From Fig. 6 it follows that the sampling RMSE for the KF is consistent with the theoretical values, and the prediction error reaches the a priori level after 6 sec of prediction. A similar result is observed for the NN, except that at the first step of the prediction, the RMSE for the NN is greater than for the KF.

From the plots it is obvious that in both cases the prediction RMSE does not exceed 6 arc min for a time of 20% of the pitching period, which corresponds to 2 sec.

6. Conclusion

The algorithms based on the Kalman filter and an artificial neural network that are used to solve the problem of pitch prediction are investigated on the simulation and real data. The sensitivity of the algorithms to model parameters changing and to the measurement noise level is analyzed.

The advantages of the algorithm based on the Kalman filter, which provides a higher accuracy, including in the presence of measurement errors, are shown. In addition, its use, in contrast to the NN-based algorithm, allows one to obtain characteristics of the prediction accuracy in real time, which is shown when processing real data.

The algorithm using the NN, in contrast to the algorithm based on the Kalman filter, showed significant sensitivity to the measurement errors. Both algorithms are sensitive to changes in model parameters. It is noted that when using real data, it is not possible to use an implementation without measurement errors as a reference during NN training. The above disadvantage leads to the presence of a prediction error already at the first step, as well as the appearance of a noise component.

In the future, it is planned to continue research in terms of the application of adaptive algorithms, including those based on the Kalman filter [15–17] and the use of more complex neural networks.

References

[1] Kalman, R. E., and Bucy, R. S. New Results in Linear Filtering and Prediction Theory. ASME. *J. Basic Eng*. March 1961; 83(1): 95–108. https://doi.org/10.1115/1.3658902

[2] Stepanov, O.A. Kalman filtering: Past and present. An outlook from Russia. (On the occasion of the 80th birthday of Rudolf Emil Kalman). *Gyroscopy Navig.* **2** 99–110 (2011)

[3] Sidar, M., & Doolin, B.F. (1983). On the feasibility of real-time prediction of aircraft carrier motion at sea. *IEEE Transactions on Automatic Control, 28*, 350-356.

[4] Chistyakov, S.P. Sluchajnye lesa: obzor. Trudy Karel'skogo nauchnogo centra Rossijskoj akademii nauk, N.1, 2013, pp. 117–136. (In Russian)

[5] Shitikov, V. K., Mastickij, S. E. (2017) Klassifikaciya, regressiya, algoritmy Data Mining s ispol'zovaniem R. E-book, access address: https://github.com/ranalytics/data-mining (In Russian)

[6] Nazarov, A.V, Loskutov, A.I Nejrosetevye algoritmy prognozirovaniya i optimizacii sistem. Spb: Nauka i tekhnika, 2003, 384 P. (In Russian)

[7] Loparev, A.V., Zaitsev, O.V., Antonov, D.V. Ispol'zovanie adaptivnyh algoritmov prognozirovaniya ugla kachki korablya. Materialy konferencii «Upravlenie v morskih sistemah» (UMS-2020). SPb.: JSC Concern Electropribor, 2020. (In Russian)

[8] Antonov, D. Comparative analysis of algorithms for ship pitching forecasting. IOP Conf.Ser.: Mater/Sci/Eng.984.012017. DOI:10.1088/1757-899X/984/1/012007

[9] Zaitsev O.V. Taking into account constraints on magnitude and first derivative of the Global Navigation Satellite Systems differential corrections during their forecasting // IOP Conf. Ser.: Mater/Sci/Eng.984.012017

[10] Stepanov, O.A. Osnovy teorii ocenivaniya s prilozheniyami k zadacham obrabotki navigacionnoj informacii. CHast' 2 Vvedenie v teoriyu fil'tracii. Sankt-Peterburg, 2017, 428p (In Russian)

[11] Lukomskij, Yu.A., Chugunov, V.S., Sistemy upravleniya morskimi podvizhnymi ob"ektami: Uchebnik – L.: Sudostroenie, 1988. 272 p. (In Russian)

[12] Haykin, S. (1999). Neural Networks: A Comprehensive Foundation. Prentice Hall.
[13] Stepanov, O.A., Amosov, O.S. Nonrecurrent linear estimation and neural networks IFAC Proceedings Volumes (IFAC-PapersOnline), 2004, 37(12), p. 213–218.
[14] Stepanov, O.A., Koshaev, D.A. A program for designing linear filtering algorithms for integrated navigation systems. IFAC Proceedings Volumes (IFAC-PapersOnline), 2011, 44(1 PART 1), p. 4256–4259
[15] Stepanov, O.A., Koshaev, D.A., Motorin, A.V. Identification of gravity anomaly model parameters in airborne gravimetry problems using nonlinear filtering methods. *Gyroscopy Navig.* **6** 318–323 (2015).
[16] Stepanov O., Motorin A. Performance criteria for the identification of inertial sensor error models/ Sensors. 2019. T. 19. № 9
[17] Stepanov, O.A., Motorin, A.V., Vasilyev, V.A. Identification of Sensor Errors by Using of Nonlinear Filtering IFAC-PapersOnLine, 2015, 48(11 11), pp. 808–813

Acknowledgments

This work was financially supported by Russian Science Foundation project № 18-19-00627.
https://rscf.ru/project/18-19-00627/
Special thanks to the research supervisor – O.A. Stepanov, Dr. Sci., corresponding member of the RAS, Elektropribor, ITMO

IOP Conf. Series: Materials Science and Engineering 1215 (2022) 012003 doi:10.1088/1757-899X/1215/1/012003

Compensation for the Temperature Dependence of the Fog Output Signal

D.A. Gontar, E.V. Dranitsyna

ITMO University, Concern CSRI Elektropribor, JSC Saint Petersburg, Russia

denisgontar99@gmail.com, dranitsyna_ev@mail.ru

Abstract. This paper proposes a method for compensating the temperature error in the FOG output signal using neural networks. One of the main advantages of the method lies in possibility for identification of complex dependencies without losing compensation accuracy at the boundaries of the temperature range.

1. Introduction

Fiber-optic gyroscopes (FOG) have been used in navigation, stabilization, and orientation systems for about thirty years. During this period, the manufacturing technologies of the FOG components have significantly improved, which allowed them to move from the class of low- and medium-precision angular velocity sensors to the class of navigation grade sensors. One of the main problems when creating high-precision FOG is the sensitivity of the output signal to temperature.

The FOG is extremely sensitive to changes in the temperature field [1]. In order to create a navigation accuracy class FOG, it is necessary to accurately compensate for the influence of temperature on the FOG output signal.

The change in temperature field affects such components of the FOG as a light source, a multifunctional integrated optical circuit, a fiber-optic coil, a processing board. Let's consider the mechanisms of influence on the components of the FOG in more detail:

- Light source: the wavelength of the source depends on the temperature
- Multifunctional integrated optical circuit: temperature changes cause inaccuracy of the coupling of the optical fiber and multifunctional integrated optical circuit, when the temperature changes, the electro-optical coefficient of the modulator changes
- Fiber-optic coil: the change in the geometric size of the fiber-optic coil when the temperature changes due to the finite value of the coefficient of linear thermal expansion of the optical fiber, the refractive index of the fiber depends on the temperature, there is a thermally induced non-reciprocity due to the thermal expansion of the fiber, a change in the refractive index due to the arising and changing thermoelastic stresses in the optical fiber.
- Processing board: changes and instability of electronic components of the board under the influence of temperature [1].

In the course of this work we are going to focus on the temperature influence on the fiber-optic coil, because it is the most vulnerable, due to the fact that in accurate FOG the fiber length reaches several kilometers.

IOP Conf. Series: Materials Science and Engineering 1215 (2022) 012003 doi:10.1088/1757-899X/1215/1/012003

To compensate for the temperature dependence of the FOG output signal we can use: linear regression models, polynomial regression models, neural networks. In the course of this paper we will be discussing compensation using neural networks.

The goal is to implement a neural network that can compensate the temperature dependence of the output signal of a FOG based on temperature, temperature gradient, and spatial distribution of the temperature field data.

2. Experimental setup

To build a model for compensation of the zero offset, caused by temperature changes and the instability of the scale factor, eight thermal sensors were placed evenly on a fiber-optic coil.

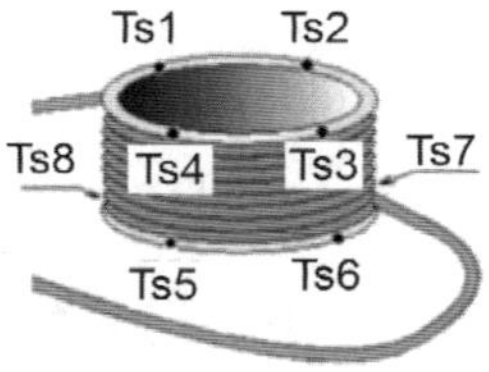

Figure 1. Diagram of the location of temperature sensors on a FOC

During the thermal calibration, the FOG was placed into a thermal camera. The temperature regime was set in the thermal chamber, which provides the maximum overlap of the temperature values and the rates of temperature change in the considered range. Temperature regime graphs are shown in Fig. 2. FOG measured the Earth's rotation speed equal to 13.0259 deg/h. Fig. 3 shows the dependence of the FOG output signal on the average temperature value from 8 sensors.

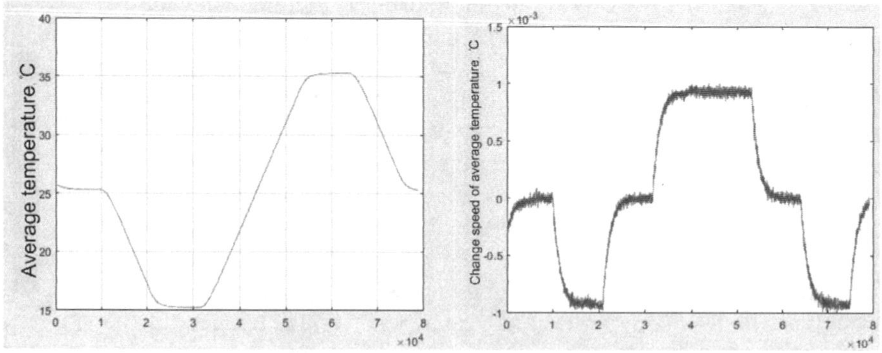

Figure 2. Temperature regime graphs

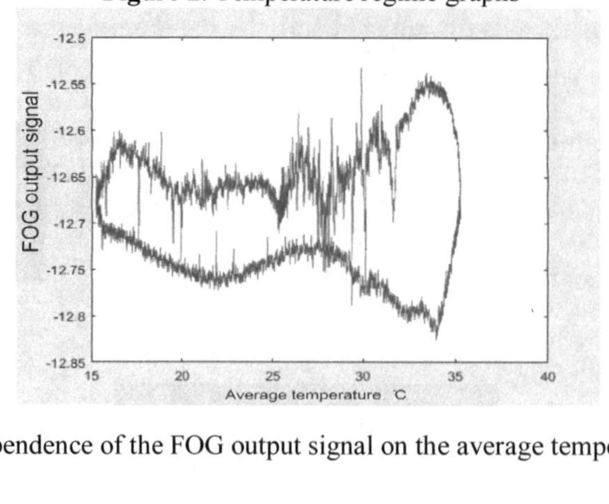

Figure 3. Graph of the dependence of the FOG output signal on the average temperature value from 8 sensors

3. Neural network structure

The Matlab software environment was chosen for simulation and training of the neural network, since this software environment provides ready-made neural networks that are easy to use, from which the user can choose a neural network suitable for his task. For this task, the "Fitting app" network was selected.

We used a two-layer feedforward network with a sigmoidal activation function of neurons of the hidden layer of neurons and a linear activation function of neurons of the output layer will be used. This network allows us to solve problems of multidimensional approximation as accurately as possible, provided that the data is consistent and the number of neurons in the hidden layer is greater than the number of inputs. The sigmoid activation function allows you to amplify weak signals and not get saturated from strong signals [2].

Our network should get the temperature from 8 temperature sensors located on the fiber-optic coil and the average temperature change rate as inputs and provide the value of temperature-educed error as an output. The mean absolute error of the FOG output values and the values predicted by the neural network difference is going to be the assessment criterion for the compensation accuracy.

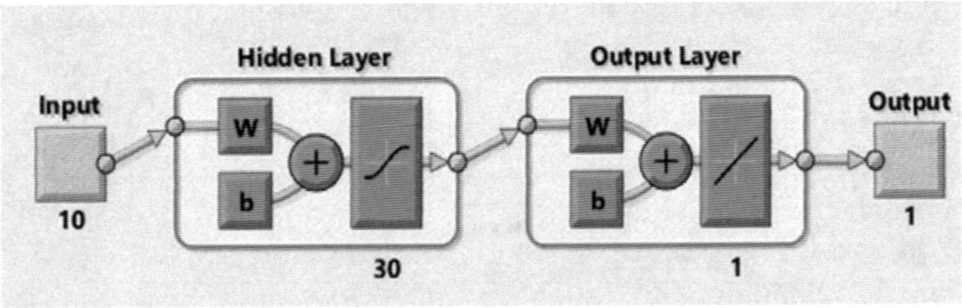

Figure 4. Generated neural network diagram

4. Neural Network Training

For the first test batch, an array consisting of temperature values from each of the 8 temperature sensors located inside the fiber-optic gyroscope housing, as well as the rate of change of the average temperature from 8 sensors, is fed to the as “Input” training data. As “Output” training data, the values of the errors of the fiber-optic gyroscope at each moment of time, calculated by subtracting from the angular velocity of the Earth's rotation, the angular velocity obtained at the output of the fiber-optic gyroscope, are fed.

In this paper we used these network training methods:

1. The method of back propagation of the error using the Levenberg-Marquardt algorithm. The error back propagation method is a method of training a neural network, the purpose of which is: to adjust each weight in proportion to how much it contributes to the overall error [3].
2. Bayesian regularization method, which updates the weight and bias values in accordance with the Levenberg-Marquardt optimization. It minimizes the combination of error squares and weights, and then determines the right combination to create a network that generalizes well [3].
3. Gradient descent is the simplest method for consistently finding the minimum of a differentiable function f (in the case of neural networks, this is a cost function). Having several parameters x (network weights) and differentiating the function by them, we obtain a vector of partial derivatives or a gradient vector. The gradient always points in the direction of the maximum growth of the function. If we move in the opposite direction (i.e. $-\nabla f(x)$) then over time we will come to a minimum, which is what we needed [3].

For the second test batch, as training data, the values of the average temperature from 8 sensors and the rate of its change are fed to the input of the neural network. Reducing the number of input parameters from 9 to 2 will greatly reduce the resource intensity of this method of temperature dependence compensation, and since the temperature difference on 8 sensors is rarely large, the error of this method should not increase much. The network is trained by the Bayesian regularization method, since tests conducted with these input data have shown that it gives the smallest decrease in accuracy.

The network trained by the Levenberg-Marquardt method with data from 8 temperature sensors, as well as the rate of change in the average temperature showed the best results.

Nevertheless, the option of using Bayesian regularization when training the network with the values of the average temperature from the sensors and the rate of its change is also justified if the device is not subjected to local temperature changes and heating and cooling of the device occurs evenly, but it is worth considering that although this method reduces the resource intensity, its accuracy is lower. In table 1 values of the errors of these networks, calculated by finding the average of the difference between the results provided by the neural network from the expected result, are shown. On figure 5 results of compensation are shown.

Table 1. The average value of the compensation error of various methods of training

Input data	Temperature, Speed of temperature change	Temperature, Speed of temperature change	Temperature, Speed of temperature change	Average temperature, Speed of temperature change	-
Training method	Levenberg-Marquardt	Bayesian regularization	Gradient descent	Bayesian regularization	Before compensation
The average value of the compensation error, deg/h	0.0050	0.0053	0.0064	0.0054	0.3387

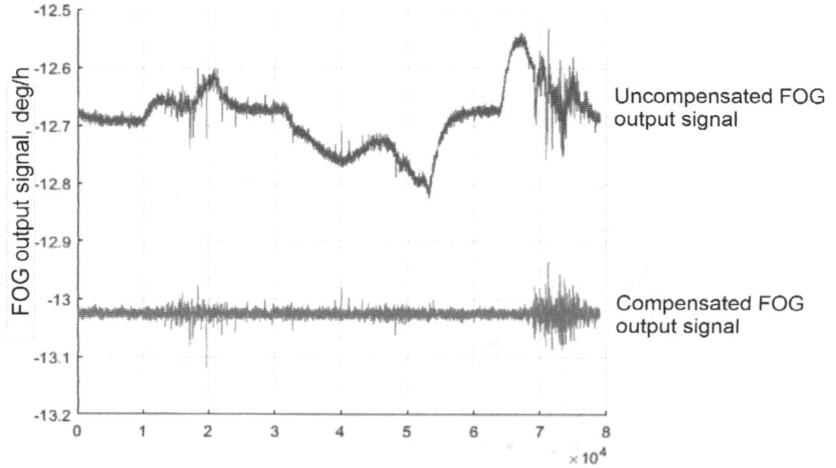

Figure 5. Compensated and uncompensated FOG output signals

5. Evaluation of temperature compensation accuracy

Now let's evaluate the temperature compensation accuracy. FOG was subjected to the temperature change that occurred in the range from -15 to 45 degrees Celsius.

Network was trained by the Levenberg-Marquardt method, and after training it was used on an unfamiliar sample. The results of the compensation are shown on the figure 6. The compensation allowed to decrease the average error by 4 times, from 0.2315 to 0.0514 deg/h.

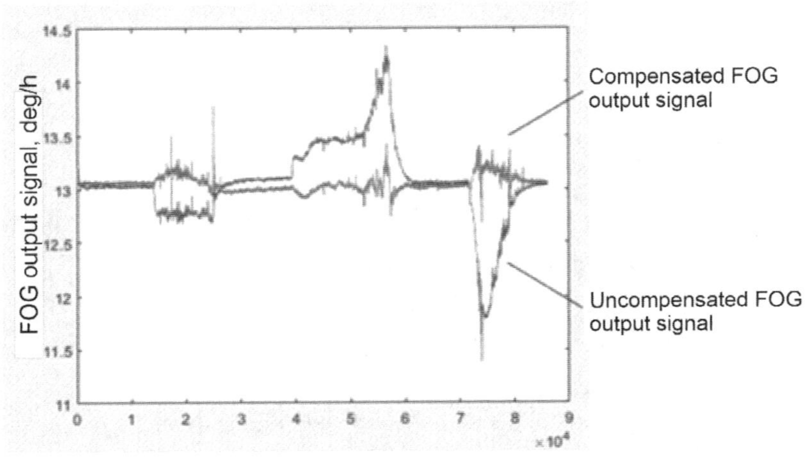

Figure 6. Compensated and uncompensated FOG output signals

6. Conclusions

The results of the work show the possibility of using neural networks to compensate the temperature dependence of the output signal of a FOG. An accuracy of $5.2*10^{-2}$ deg/h was achieved in the operating temperature range of -15-45 °C, which is four times better than before compensation. The algorithms, the structure of neural networks given in this paper can be used in systems with sufficient computing power to accurately compensate for the temperature dependence of the output signal of a fiber-optic gyroscope.

References

[1] E.V. Dranitsyna, D.A. Egorov. Investigation of the dependence of the output signal of a fiber-optic gyroscope on temperature as part of a cardeless inertial measuring module / XIV Conference of young scientists "Navigation and motion control". St. Petersburg: JSC "Concern "Central Research Institute "Electropribor". 2012. - pp. 447-452. (in Russian)

[2] Neural networks for beginners. Part 1. [Electronic resource] // HABR - 2016. - Access mode: Neural networks for beginners. Part 1 / Habr (habr.com). (in Russian)

[3] Goodfellow Ya., Benjio I., Courville A. Deep learning / 2nd ed. M.: DMK Press, 2018. 652 p.

[4] Klimkovich, B.V. Effect of Random Error of Temperature Sensors on the Quality of Temperature Compensation of FOG Bias by a Neural Network. *Gyroscopy Navig.* **12**, 27–37 (2021). https://doi.org/10.1134/S2075108721010089

[5] Gorski, P., Ledzion, R., Bondarczuk, K., Kucharczyk, W. Temperature dependence of linear electrooptic coefficients r113 and r333 in lithium niobate. *Opto-Electronics Review.* March 2008, Volume **16**, Issue 1, pp. 46-48. (in Russian)

[6] Mohr, F., Schadt, F. Error signal formation in FOGs through thermal and elastooptical environ-mental influences on the sensing coil. // *Inertial Sensors and Systems*, 2011. C. 2.1-2.13

[7] Golikov, A.V. Temperature errors of fiber-optic gyroscopes: Dis. ... Candidate of Technical Sciences. Saratov. 2001. - 197 p. (in Russian)

[8] Dranitsyna, E.V., Egorov, D.A., Untilov, A.A. et al. Reducing the effect of temperature variations on FOG output signal. *Gyroscopy Navig.* **4**, 92–98 (2013). https://doi.org/10.1134/S2075108713020041

[9] Meshkovsky, I.K. Investigation of the influence of thermal influences on the operation of a fiber-optic angular velocity sensor / I.K. Meshkovsky [et al.] // XXI St. Petersburg International Conference on Integrated Navigation Systems: Collection of Materials - St. Petersburg, - 2014. - pp.191 - 202. 51. (in Russian)

[10] Neural network [Electronic resource]//Wikipedia - 2021. - Access mode: Neural Network - Wikipedia (wikipedia.org)

Calibration of the Azimuth Drive Angle Sensor of the Mobile Automated Astronomical System by Stars Observation

S. M. Tarasov

Concern CSRI Electropribor, JSC, Saint Petersburg, Russia

sergione@list.ru

Abstract. The paper describes the design of the mobile automated astronomical system (MAAS) and the technique of the MAAS azimuth drive calibration by the zenithal stars observation. This technique provides fast and accurate calibration of the MAAS azimuth drive angle sensor in field conditions.

1. Introduction

Devices for the astronomical azimuth determination are used in geodesy and survey operations in field conditions [1]. Gyrocompasses and gyrotheodolites are commonly used for determination of the objects attitude [2]. The proper operation of such devices needs calibration to reduce the systematic error of the azimuth determination [3].

Calibration of gyrocompasses and gyrotheodolites performs with base directions, which azimuths are known with high accuracy. Base directions are fixed by reference marks and reflecting devices. In field conditions azimuths of the base directions are determines with classical astronomical theodolites [1-4] and total stations [5, 6].

The mobile automated astronomical system (MAAS) is being developed for precise and fast azimuth and coordinates determination in field conditions [7]. It is supposed to substitute the classical astronomical theodolites.

The MAAS is designed as a classical astronomical theodolite and comprising several main parts (Fig. 1):

- sighting device (SD) consisting of the objective and photodetector;
- vertical drive (VD) and azimuthal drive (AD) for rotating the SD sighting axis in the vertical and horizontal plane;
- control panel (PC) and GPS/GLONASS receiver.

Design of the AD and VD based on torque motors and digital angle sensors. The drives operate in automatic mode via PC control commands.

The MAAS also comprises a two-axis inclination sensor (IS) installed on AD rotating part and a leveling system for setting the AD rotation axis vertical to the horizontal plane.

The MAAS determines the landmark azimuth as follows: it determines the azimuth of sighting axis A_{SA} by stellar observations near the meridian plane and then turns it to the landmark, with the horizontal rotation angle of sighting axis γ calculated by AD angle sensor data. Thus the astronomical azimuth of a landmark is determined by the formula:

$$A_L = A_{SA} + \gamma . \quad (1)$$

IOP Conf. Series: Materials Science and Engineering 1215 (2022) 012004 doi:10.1088/1757-899X/1215/1/012004

It is clear that the error of the rotation angle γ affect on the error of the landmark azimuth determination. The error of the rotation angle γ includes the error of the angle sensor [7].

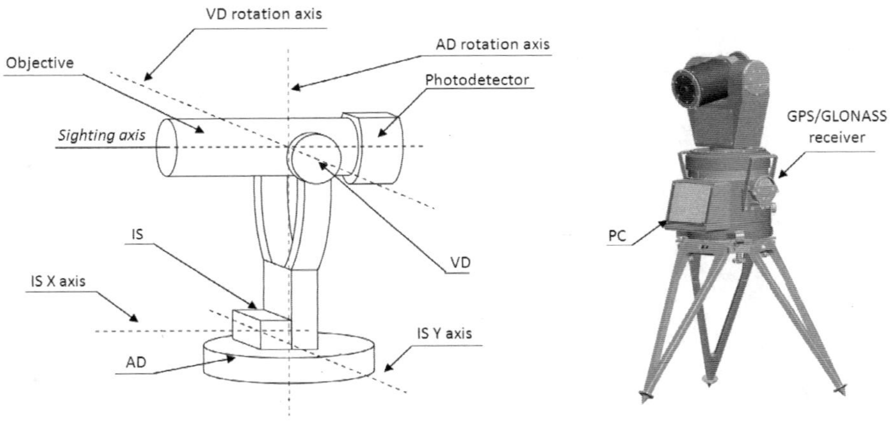

Figure 1. Mobile automated astronomical system

The most precise angle sensors are optical encoders LIR-4200 (SKBIS, JSC, Russia) and RON-905 (Heidenhain, Germany) with limits of errors ±0,4 arcsec [8]. These sensors are not suitable for the MAAS AD because of high temperature sensitivity (temperature operation conditions of the MAAS are in the range from minus 30 to 40°C). The MAAS AD includes the temperature independent sensor of another type whose limit of error is ±1,5 arcsec. High accuracy of the angle sensor is provided by postmanufacturing calibration after in factory conditions [9].

However, calibration of the angle sensor in field conditions is still needed before astrogeodetic operations with the MAAS because of several factors: errors of the angle sensor installation into the MAAS azimuth drive, thermal deformations of the MAAS, and vibrations during transportation.

This paper describes the technique of the MAAS azimuth drive angle sensor calibration by the zenithal stars observation.

2. The technique of the AD angle sensor calibration by the zenithal stars observation

The problem of determining the calibration parameters of the AD angle sensor is solved by comparing the sensor measured angle with the reference value of this angle obtained with a more accurate angle measuring device [10].

Determination of the reference horizontal rotation angle around the AD rotation axis can be realized by calculating the rotation angle of the image, which contains number of point objects with the definite positions [11].

It is proposed to use images of stars in the near-zenith region as a number of point objects with definite positions for the MAAS AD angle sensor calibration.

Before calibration, the SD sighting axis set up in position for zenithal stars observation (vertical to the horizontal plane).

The method of calibration of the MAAS AD angle sensor consists in performing consecutive turns of the SD around the AD rotation axis by the angle β_j, while the following actions are performed at each j-th turn:

1) the reference rotation angle $\beta_{\mathrm{Row}j}$ (the SD angle of rotation around the AD rotation axis) is determined using astronomical observation data in accordance with the algorithm:
   (a) in the position before turn by the angle β_j, N_f frames containing N_s images of stars are recorded by the photodetector. Then arrays of the star image coordinates $\{x^{\mathrm{I}}_{1k} \ldots x^{\mathrm{I}}_{Nsk}\}$ and $\{y^{\mathrm{I}}_{1k} \ldots y^{\mathrm{I}}_{Nsk}\}$, where k=1…N_f, and arrays of equatorial coordinates of these stars $\{\alpha^{\mathrm{I}}_1 \ldots \alpha^{\mathrm{I}}_{Ns}\}$ и $\{\delta^{\mathrm{I}}_1 \ldots \delta^{\mathrm{I}}_{Ns}\}$ are formed;

(b) for each k-th frame, the astronomical azimuth of the photodetector row $A^{\mathrm{I}}_{\mathrm{Row}k}$ is determined by the next stages:

- the system of equations connecting the photodetector and equatorial coordinate systems is formed in accordance with the expressions

$$\begin{cases} \xi_1 = C_{11} + C_{12} x^{\mathrm{I}}_{1k} + C_{13} y^{\mathrm{I}}_{1k}; \\ \xi_2 = C_{11} + C_{12} x^{\mathrm{I}}_{2k} + C_{13} y^{\mathrm{I}}_{2k}; \\ \dots \\ \xi_{Ns} = C_{11} + C_{12} x^{\mathrm{I}}_{Nsk} + C_{13} y^{\mathrm{I}}_{Nsk}. \end{cases} \quad \begin{cases} \eta_1 = C_{21} + C_{22} x^{\mathrm{I}}_{1k} + C_{23} y^{\mathrm{I}}_{1k}; \\ \eta_2 = C_{21} + C_{22} x^{\mathrm{I}}_{2k} + C_{23} y^{\mathrm{I}}_{2k}; \\ \dots \\ \eta_{Ns} = C_{21} + C_{22} x^{\mathrm{I}}_{Nsk} + C_{23} y^{\mathrm{I}}_{Nsk}. \end{cases}, \quad (2)$$

where C_{11}, C_{12}, C_{13}, C_{21}, C_{22}, C_{23} – the parameters of the stars coordinates transformation into their orthogonal tangent coordinates associated with the equatorial coordinate system;

ξ_i, η_i – orthogonal tangent coordinates of the stars, that are calculated by equatorial coordinates of the stars in accordance with the expressions [12]

$$\xi_i = \frac{\mathrm{ctg}(\delta^{\mathrm{I}}_i)\sin(\alpha^{\mathrm{I}}_i - \alpha_0)}{\sin(\delta_0) + \mathrm{ctg}(\delta^{\mathrm{I}}_i)\cos(\delta_0)\cos(\alpha^{\mathrm{I}}_i - \alpha_0)}; \quad \eta_i = \frac{\cos(\delta_0) - \mathrm{ctg}(\delta^{\mathrm{I}}_i)\sin(\delta_0)\cos(\alpha^{\mathrm{I}}_i - \alpha_0)}{\sin(\delta_0) + \mathrm{ctg}(\delta^{\mathrm{I}}_i)\cos(\delta_0)\cos(\alpha^{\mathrm{I}}_i - \alpha_0)}, \quad (3)$$

where α^{I}_i, δ^{I}_i – the equatorial coordinates of the i-th star;
α_0, δ_0 – the equatorial coordinates of the visible area center;
i=1…N_s;

- the system of equations (2) is solved by the least squares method:

$$C_1 = (H^T H)^{-1} H^T \cdot \xi \; ; \; C_2 = (H^T H)^{-1} H^T \cdot \eta, \quad (4)$$

where

$$\xi = \begin{bmatrix} \xi_1 \\ \xi_2 \\ \vdots \\ \xi_{Ns} \end{bmatrix}; \eta = \begin{bmatrix} \eta_1 \\ \eta_2 \\ \vdots \\ \eta_{Ns} \end{bmatrix}; H = \begin{bmatrix} 1 & x^{\mathrm{I}}_{1k} & y^{\mathrm{I}}_{1k} \\ 1 & x^{\mathrm{I}}_{2k} & y^{\mathrm{I}}_{2k} \\ \vdots & \vdots & \vdots \\ 1 & x^{\mathrm{I}}_{Nsk} & y^{\mathrm{I}}_{Nsk} \end{bmatrix}; C_1 = \begin{bmatrix} C_{11} \\ C_{12} \\ C_{13} \end{bmatrix}; C_2 = \begin{bmatrix} C_{21} \\ C_{22} \\ C_{23} \end{bmatrix}. \quad (5)$$

- astronomic azimuth of the photodetector row $A^{\mathrm{I}}_{\mathrm{Row}k}$ is calculated in accordance with the expression

$$A^{\mathrm{I}}_{\mathrm{Row}k} = \mathrm{atan}\left(\frac{C_{13}}{C_{23}}\right). \quad (6)$$

(c) the average photodetector row azimuth value $A^{\mathrm{I}}_{\mathrm{Row}}$ is calculated using the equation

$$A^{\mathrm{I}}_{\mathrm{Row}} = \frac{\sum_{k=1}^{N\mathrm{f}} A^{\mathrm{I}}_{\mathrm{Row}k}}{N_{\mathrm{f}}}. \quad (7)$$

(d) the SD is rotated around the AD rotation axis by an angle β_j, after which N_{f} frames containing N_s stars are recorded again. Then arrays of the star image coordinates $\{x^{\mathrm{II}}_{1k} \dots x^{\mathrm{II}}_{Nsk}\}$, $\{y^{\mathrm{II}}_{1k} \dots y^{\mathrm{II}}_{Nsk}\}$ and arrays of equatorial coordinates of these stars $\{\alpha^{\mathrm{II}}_1 \dots \alpha^{\mathrm{II}}_{Ns}\}$, $\{\delta^{\mathrm{II}}_1 \dots \delta^{\mathrm{II}}_{Ns}\}$ are formed;

(e) the astronomical azimuth of the photodetector row $A^{\mathrm{II}}_{\mathrm{Row}}$ in after-turn position is calculated according to steps a) – c);

(f) the angle of rotation around the AD rotation axis $\beta_{\mathrm{Row}j}$ is calculated in accordance with the expression

$$\beta_{\mathrm{Row}j} = A^{\mathrm{II}}{}_{\mathrm{Row}} - A^{\mathrm{I}}{}_{\mathrm{Row}}; \tag{8}$$

2) the angle of rotation around the AD rotation axis $\beta_{\mathrm{AD}j}$ measured by the AD angle sensor is calculated using the formula

$$\beta_{\mathrm{AD}j} = \beta^{\mathrm{II}}{}_{\mathrm{AD}j} - \beta^{\mathrm{I}}{}_{\mathrm{AD}j}, \tag{9}$$

where $\beta^{\mathrm{I}}{}_{\mathrm{AD}j}$, $\beta^{\mathrm{II}}{}_{\mathrm{AD}j}$ – the AD angle sensor readings before and after turning by the angle β_j, respectively;

3) the correction $\Delta_{\mathrm{AD}j}$ for the AD angle sensor readings is calculated in accordance with the expression

$$\Delta_{\mathrm{AD}j} = \beta_{\mathrm{Row}j} - \beta_{\mathrm{AD}j}. \tag{10}$$

3. Study of the error in determining the SD rotation angle around the AD rotation axis

3.1. Simulation of the astronomical observation process

The estimation of the error in determining the angle of rotation of the SD around the AD rotation axis was performed by simulating in the Matlab software environment.

As part of the simulations, the astronomical observation process was simulated according to the next steps:

(a) N_{f} frames containing N_{s} images of stars were formed. For each of star the coordinates in the photodetector plane $\{x^{\mathrm{I}}{}_{1k} \ldots x^{\mathrm{I}}{}_{Nsk}\}$ and $\{y^{\mathrm{I}}{}_{1k} \ldots y^{\mathrm{I}}{}_{Nsk}\}$, the equatorial coordinates $\{\alpha^{\mathrm{I}}{}_1 \ldots \alpha^{\mathrm{I}}{}_{Ns}\}$ and $\{\delta^{\mathrm{I}}{}_1 \ldots \delta^{\mathrm{I}}{}_{Ns}\}$, and their orthogonal tangent coordinates $\{\xi^{\mathrm{I}}{}_{1k} \ldots \xi^{\mathrm{I}}{}_{Nsk}\}$ and $\{\eta^{\mathrm{I}}{}_{1k} \ldots \eta^{\mathrm{I}}{}_{Nsk}\}$ are known.

When modeling the first frame, the image star coordinates $\{x^{\mathrm{I}}{}_{11} \ldots x^{\mathrm{I}}{}_{Ns1}\}$ and $\{y^{\mathrm{I}}{}_{11} \ldots y^{\mathrm{I}}{}_{Ns1}\}$ were formed randomly. To calculate the orthogonal tangent coordinates $\xi^{\mathrm{I}}{}_{i1}$, $\eta^{\mathrm{I}}{}_{i1}$ of the simulated stars, the transformation parameters were set

$$C^{\mathrm{s}}{}_{11}=0;\ C^{\mathrm{s}}{}_{12}=M\cdot\cos(\gamma);\ C^{\mathrm{s}}{}_{13}=-M\cdot\sin(\gamma); \tag{11}$$
$$C^{\mathrm{s}}{}_{21}=0;\ C^{\mathrm{s}}{}_{22}=M\cdot\sin(\gamma);\ C^{\mathrm{s}}{}_{23}=M\cdot\cos(\gamma),$$

where γ – the rotation angle of the photodetector coordinate system around equatorial coordinate system. The angle γ in the simulation varied from 0 to 360° with increment $\beta_j = 10°$;
M – the image scale factor. In the simulation $M = 2$ arcsec/pixel.

Orthogonal tangent coordinates $\xi^{\mathrm{I}}{}_{i1}$, $\eta^{\mathrm{I}}{}_{i1}$ of the stars were calculated using the formulas

$$\xi^{\mathrm{I}}{}_{i1}= C^{\mathrm{s}}{}_{11}+ x^{\mathrm{I}}{}_{i1}\cdot C^{\mathrm{s}}{}_{12}+ y^{\mathrm{I}}{}_{i1}\cdot C^{\mathrm{s}}{}_{13}; \tag{12}$$
$$\eta^{\mathrm{I}}{}_{i1} = C^{\mathrm{s}}{}_{21}+ x^{\mathrm{I}}{}_{i1}\cdot C^{\mathrm{s}}{}_{22}+ y^{\mathrm{I}}{}_{i1}\cdot C^{\mathrm{s}}{}_{23},$$

after which the equatorial coordinates of the simulated stars $\alpha^{\mathrm{I}}{}_i$, $\delta^{\mathrm{I}}{}_i$ were calculated according to the equations [12]

$$\alpha^{\mathrm{I}}_i = \alpha^{\mathrm{s}}_0 + \mathrm{atan}\left(\frac{\xi^{\mathrm{I}}_{i1}}{\cos\delta^{\mathrm{s}}_0 - \eta^{\mathrm{I}}_{i1}\sin\delta^{\mathrm{s}}_0}\right);\ \delta^{\mathrm{I}}_i = \mathrm{atan}\left(\frac{\left(\eta^{\mathrm{I}}_{i1} + \mathrm{tg}\delta^{\mathrm{s}}_0\right)\cos(\alpha^{\mathrm{I}}_i - \alpha^{\mathrm{s}}_0)}{1-\eta^{\mathrm{I}}_{i1}\mathrm{tg}\delta^{\mathrm{s}}_0}\right), \tag{13}$$

where $\alpha^{\mathrm{s}}{}_0$, $\delta^{\mathrm{s}}{}_0$ – simulated equatorial coordinates of the visible area center. In the simulation $\alpha^{\mathrm{s}}{}_0$=30°, $\delta^{\mathrm{s}}{}_0$=60°.

When modeling next frames ($k > 1$), the following actions were performed to simulate the movement of images of stars in the photodetector plane caused by the diurnal rotation of the Earth:

- similarly to the formulas (4) – (5), the inverse transformation parameters were calculated by solving the system of equations:

$$\begin{cases} x_{11}^{\mathrm{I}} = C_{11}^{\mathrm{inv}} + C_{12}^{\mathrm{inv}}\xi_{11}^{\mathrm{I}} + C_{13}^{\mathrm{inv}}\eta_{11}^{\mathrm{I}}; \\ x_{21}^{\mathrm{I}} = C_{11}^{\mathrm{inv}} + C_{12}^{\mathrm{inv}}\xi_{21}^{\mathrm{I}} + C_{13}^{\mathrm{inv}}\eta_{21}^{\mathrm{I}}; \\ \ldots \\ x_{Ns1}^{\mathrm{I}} = C_{11}^{\mathrm{inv}} + C_{12}^{\mathrm{inv}}\xi_{Ns1}^{\mathrm{I}} + C_{13}^{\mathrm{inv}}\eta_{Ns1}^{\mathrm{I}}. \end{cases} \quad \begin{cases} y_{11}^{\mathrm{I}} = C_{21}^{\mathrm{inv}} + C_{22}^{\mathrm{inv}}\xi_{11}^{\mathrm{I}} + C_{23}^{\mathrm{inv}}\eta_{11}^{\mathrm{I}}; \\ y_{21}^{\mathrm{I}} = C_{21}^{\mathrm{inv}} + C_{22}^{\mathrm{inv}}\xi_{21}^{\mathrm{I}} + C_{23}^{\mathrm{inv}}\eta_{21}^{\mathrm{I}}; \\ \ldots \\ y_{Ns1}^{\mathrm{I}} = C_{21}^{\mathrm{inv}} + C_{22}^{\mathrm{inv}}\xi_{Ns1}^{\mathrm{I}} + C_{23}^{\mathrm{inv}}\eta_{Ns1}^{\mathrm{I}}. \end{cases} \tag{14}$$

- using the formula (3), the orthogonal tangent coordinates of the stars in the k-th frame were calculated. The equatorial coordinates of the center of the visible area α_{0k}, δ_{0k} were set in accordance with the formulas

$$\delta_{0k}=\delta^{s}_{0},\ \ \alpha_{0k}=\alpha^{s}_{0}+\omega_{E}\cdot\cos(\delta^{s}_{0})\cdot(k\text{-}1), \tag{15}$$

where ω_E – the angular velocity of the diurnal rotation of the Earth. In the simulation ω_E=15 arcsec/s.

- using the inverse transformation parameters $C_{11}^{\mathrm{inv}}, C_{12}^{\mathrm{inv}}, C_{13}^{\mathrm{inv}}, C_{21}^{\mathrm{inv}}, C_{22}^{\mathrm{inv}}, C_{23}^{\mathrm{inv}}$, the star image coordinates x^{I}_{ik}, y^{I}_{ik} shifted due to the diurnal rotation of the Earth was calculated by the formulas

$$x_{ik}^{\mathrm{I}} = C_{11}^{\mathrm{inv}} + C_{12}^{\mathrm{inv}}\xi_{ik}^{\mathrm{I}} + C_{13}^{\mathrm{inv}}\eta_{ik}^{\mathrm{I}};\ \ y_{ik}^{\mathrm{I}} = C_{21}^{\mathrm{inv}} + C_{22}^{\mathrm{inv}}\xi_{ik}^{\mathrm{I}} + C_{23}^{\mathrm{inv}}\eta_{ik}^{\mathrm{I}}; \tag{16}$$

(b) the errors $\Delta\alpha_i^{\mathrm{I}}, \Delta\delta_i^{\mathrm{I}}$, in determining the equatorial coordinates in the star catalogue that are characterized by the standard deviation $\sigma_{\alpha\delta}$ = 0,05 arcsec and zero conditional expectations were simulated and then added to the stars equatorial coordinates using the formulas

$$\alpha_i^{\mathrm{Ierr}} = \alpha_{ik}^{\mathrm{I}} + \Delta\alpha_i^{\mathrm{I}};\ \ \delta_i^{\mathrm{Ierr}} = \delta_{ik}^{\mathrm{I}} + \Delta\delta_i^{\mathrm{I}}; \tag{17}$$

(c) for each i-th star in k-th frame the errors $\Delta x_{ik}^{\mathrm{I}}, \Delta y_{ik}^{\mathrm{I}}$ in determining the star image coordinates that are characterized by the standard deviation σ_{xy} and zero conditional expectations were simulated and then added to the star image coordinates according to the equations

$$x_{ik}^{\mathrm{Ierr}} = x_{ik}^{\mathrm{I}} + \Delta x_{ik}^{\mathrm{I}};\ \ y_{ik}^{\mathrm{Ierr}} = y_{ik}^{\mathrm{I}} + \Delta y_{ik}^{\mathrm{I}}. \tag{18}$$

In the simulation the standard deviation σ_{xy} was varied from 0,1 to 0,2 pixel;

(d) the errors $\Delta\alpha_{0k}, \Delta\delta_{0k}$ in determining the equatorial coordinates of the visible area center that are characterized by the standard deviations $\sigma_{\alpha0}$=0,45 arcsec, $\sigma_{\delta0}$=0,3 arcsec and zero conditional expectations were calculated and then added to the simulated equatorial coordinates of the visible area center using the formulas

$$\alpha_{0k}^{\mathrm{err}} = \alpha_{0k} + \Delta\alpha_{0k};\ \ \delta_{0k}^{\mathrm{err}} = \delta_{0k} + \Delta\delta_{0k}; \tag{19}$$

(e) using the data in the arrays $\{\alpha^{\mathrm{Ierr}}_{1} \ldots \alpha^{\mathrm{Ierr}}_{Ns}\}$, $\{\delta^{\mathrm{Ierr}}_{1} \ldots \delta^{\mathrm{Ierr}}_{Ns}\}$ and the visible area coordinates $\alpha^{\mathrm{err}}_{0k}$, $\delta^{\mathrm{err}}_{0k}$, the orthogonal tangent coordinates ξ^{Ierr}_{ik}, $\eta^{\mathrm{Ierr}}_{ik}$ of the i-th star were calculated again according to formulas (3);

(f) using the data in the arrays $\{x^{\mathrm{Ierr}}_{1k} \ldots x^{\mathrm{Ierr}}_{Nsk}\}$, $\{y^{\mathrm{Ierr}}_{1k} \ldots y^{\mathrm{Ierr}}_{Nsk}\}$ and the orthogonal tangent coordinates of the stars ξ^{Ierr}_{ik}, $\eta^{\mathrm{Ierr}}_{ik}$, the average value of the photodetector row azimuth $A^{\mathrm{I}}_{\mathrm{Row}}$ before the rotation was calculated using formula (7);

(g) the SD rotation by the angle β_j was simulated by incrementing of the angle γ by β_j. Stage a) was repeated and, as a result, the new set of N_f frames with N_s images of stars were formed. For each of simulated stars the coordinates in the photodetector plane $\{x^{II}_{1k} \ldots x^{II}_{Nsk}\}$, $\{y^{II}_{1k} \ldots y^{II}_{Nsk}\}$, the equatorial coordinates $\{\alpha^{II}_1 \ldots \alpha^{II}_{Ns}\}$, $\{\delta^{II}_1 \ldots \delta^{II}_{Ns}\}$, and their orthogonal tangent coordinates $\{\xi^{II}_{1k} \ldots \xi^{II}_{Nsk}\}$ and $\{\eta^{II}_{1k} \ldots \eta^{II}_{Nsk}\}$ are known;

(h) similarly to stages b) – e), the errors of the star image coordinates, equatorial and orthogonal tangent coordinates were simulated. Then arrays of coordinates $\{x^{IIerr}_{1k} \ldots x^{IIerr}_{Nsk}\}$, $\{y^{IIerr}_{1k} \ldots y^{IIerr}_{Nsk}\}$, $\{\alpha^{IIerr}_1 \ldots \alpha^{IIerr}_{Ns}\}$, $\{\delta^{IIerr}_1 \ldots \delta^{IIerr}_{Ns}\}$, $\{\xi^{IIerr}_{1k} \ldots \xi^{IIerr}_{Nsk}\}$, $\{\eta^{IIerr}_{1k} \ldots \eta^{IIerr}_{Nsk}\}$, the average value of the photodetector row azimuth A^{II}_{Row} after rotation were calculated;

(i) using formula (8), the SD angle of rotation around the AD rotation axis β_{Rowj} was calculated;

(j) the error of the rotation angle determination $\Delta\beta_j$ was calculated according to the equation

$$\Delta\beta_j = \beta_j - \beta_{Rowj}; \tag{20}$$

(k) actions a) - k) were performed 100 times for different values of N_s in the range from 10 to 200 stars and errors in determining the star image coordinates σ_{xy}.

In the simulation the number of frames $N_к = 100$.

3.2. The results of the simulation

Fig. 2 presents the results of simulation. The graphs show the dependence of the standard error of the determination of the angle β_{Row} on the number of stars N_s and various values of star image coordinates determination errors σ_{xy}.

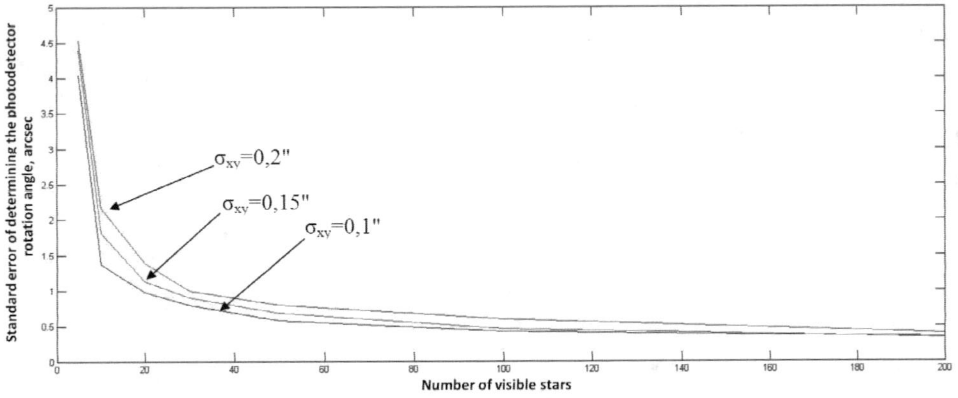

Figure 2. The dependence of the standard error in determining the angle β_{Row} on the number of stars and various values of star image coordinates determination errors

The results of the simulation show that the errors in determining the SD rotation angle around AD rotation axis and corrections $\Delta\beta_j$ for the AD angle sensor's readings does not exceed 0.5 arcsec while good observation conditions (σ_{xy}=0,1 pixel) and at least 60 stars are observed. When the observation conditions worsen (σ_{xy}> 0,1 pixel), to provide the same accuracy of determining the corrections $\Delta\beta_j$, it is necessary to observe a larger number of stars.

4. Conclusion

The paper proposes the technique for calibrating the azimuth drive angle sensor of the mobile automated astronomical system by the zenithal stars observation. The presented simulation results show the high efficiency of the calibration technique. In good observation conditions, standard error in determining the sighting device rotation angle around the azimuth drive rotation axis does not exceed 0.5 arcsec while at least 60 stars are observed.

References

[1] Sugar, D., Brkic, M. and Spoljaric, D. (2012). Comparison of the reference mark azimuth determination methods. *Annals of Geophysics*, Vol. 55, No. 6: 1071-1083.
[2] Torge, W. and Muller, J. Geodesy, Fourth edition, Berlin, Boston: De Gruyter, 2012.
[3] Glazunov, A.S., Modern trends in astrogeodesy, *Interexpo Geo-Siberia*, Novosibirsk: SGGA, 2008, pp. 183–188.
[4] Chernov, I.V., A procedure for determining the gyrotheodolite instrumental error with a preset accuracy, *Trudy voenno-kosmicheskoi akademii im. A.F. Mozhaiskogo* (Proceedings of the Mozhaiskiy Military Space Academy), St. Petersburg: Mozhaiskiy Military Space Academy, 2019, No. 671, pp. 192–200.
[5] Hauk, M., Hirt, C., Ackermann, C., Experiences with the QDaedalus system for astrogeodetic determination of deflections of the vertical, Survey Review, 2016, vol.49, pp. 294–301.
[6] Baladimos, D.D., Korakitis, R., Lambrou, E., Pantazis, G., Fast and accurate determination of astronomical coordinates Φ, Λ and azimuth using a total station and GPS receiver, Survey Review, 2003, vol. 37, pp. 269–275.
[7] Tarasov, S.M. A study on the effect produced by instrumental error of automated astronomical system on landmark azimuth accuracy. *Gyroscopy Navig.* **12**. 178-185 (2021). https://doi.org/10.1134/S2075108721020085
[8] Bokhman, E., Ivanov, P., Larichev, R., Nikolaev, M., Pavlov, P., Filatov, Yuri, Berkovich, S., Kotov, N.. Automated angle measurement system for directions transmission. *Gyroscopy Navig.* **10**. 180–185 (2019). https://doi.org/10.1134/S2075108719030039.
[9] Kir'yanov, V.P. (2005). High-precision control systems for angle-measuring structures on transparent disks // *Interexpo Geo-Siberia*, Novosibirsk: SGGA, 81–90.
[10] Kupriyanov, V.V., Devyatkin A.B. (2012). Calibration of the angle-measuring devices by CCD-observation of the stars images // Processings of the Higher Educational Institutions. Izvestia vuzov «Geodesy and aerophotosurveying». No 4. 8–15.
[11] Korolev, A., Lukin, A., Polishchuk, G. (2012). New concept of angular measurement. Model and experimental studies. *Journal of Optical Technology*. **79**. https://doi.org/10.1364/JOT.79.000352
[12] *Modern Technologies and Methods for Measuring the Earth's Gravity Field*, Peshekhonov, V.G., Stepanov, O.A., Eds., St. Petersburg: Concern CSRI Elektropribor, JSC, in press.

IOP Conf. Series: Materials Science and Engineering 1215 (2022) 012005 doi:10.1088/1757-899X/1215/1/012005

Providing Nanosatellite Triaxial Gravitational Orientation Using Magnetic Actuators

A.I. Leonov, P.N. Nikolaev

Samara National Research University, Samara, Russia

aleksey.l.i.1996@mail.ru

Abstract. Three-axis gravity stabilization of 3U CubeSat is achieved due to selection of the nanosatellite moments of inertia at the design stage, as well as special modes included in the algorithm to provide stabilization of CubeSat relative to each motion channel separately. In this paper, we propose a modified algorithm based on the magnetic stabilization algorithm B-dot. The modified algorithm provides three modes intended to damp the initial angular velocity to the value of the orbital angular velocity, to keep the angular velocity at a value close to that of the orbital angular velocity, and to provide the nanosatellite gravitational triaxial stabilization by using one magnetic coil located on the axis with the transversal moment of inertia, which is possible due to the small angle between the magnetic field line and the satellite's trajectory. We propose two modifications for forming a control loop for orientation and stabilization of the 3U CubeSat: the first one uses measurements from magnetometers and angular rate sensors as feedback, and the second one, only magnetometers. The efficiency of the two modifications of modifications was studied by means of statistical modeling.

1. Introduction

When nanosatellites carry out scientific missions, their oriented motion in space is often required. Taking into account the external aspects of the environment, at the design stage it is possible to provide such an impact of external forces moments, which helps to passively stabilize nanosatellite in a certain orientation. A special case is the triaxial gravitational orientation, which is necessary in missions that require the longitudinal axis along radius vector of nanosatellite as well as a certain orientation of the transverse axes. The gravitational moment tends to orient a nanosatellite so that the axis of the smallest moment of inertia coincides with the local vertical, the axis of the largest moment of inertia coincides with the orbital velocity vector, and the axis of the principal central moment of inertia with an intermediate value coincides with the direction of motion. In work [1] authors show an approach for forming the design parameters of nanosatellite at the development stage. These parameters are the best to form moments, supporting to main attitude determination and control system (ADCS).

Among different types of ADCS, magnetic one is extremely popular to solve the problems offered by majority of nanosatellite missions: about 40% of nanosatellites have magnetic torquers for active control [2]. Besides magnetic torquers, nanosatellite ADCS are often supplemented with various types of sensors, the most popular of which are sun sensors and magnetometers [3].

The operation of magnetic ADCS is based on the principle of interaction of an external magnetic field with a magnetic field generated by the magnetic torquers [4-7]. The specificity of this interaction is such that it is impossible at each moment of time to form the required mechanical moment, and therefore, to give to the satellite the required orientation [8]. A satellite with only a magnetic ADCS

IOP Conf. Series: Materials Science and Engineering 1215 (2022) 012005 doi:10.1088/1757-899X/1215/1/012005

have not full controllability [9], nevertheless full controllability can be achieved taking into account a slow change in the geomagnetic vector, which forms a closed control loop over the orbital period [10].

The classical implementation of ADCS in task of damping of angular velocity is B-dot control algorithm [11-14]. Nowadays there are various modifications of B-dot, for instance classical B-dot supplemented by flywheel [15]. In the work [16] authors considered another modification – Y-dot algorithm, which use just one magnetic coil. This algorithm brings the spacecraft to an equilibrium position in which it rotates about one of the axes at a constant speed depending on the initial conditions [17]. In the work [18] the magnetic control without attitude determination for spinning spacecraft is considered.

In the current work, two modifications of the B-dot control algorithm are proposed. These modifications bring the CubeSat 3U to a triaxial gravitational orientation. The study of the developed modification is carried out on the example of the SamSat-ION nanosatellite using the method of statistical modeling, which makes it possible to evaluate the effectiveness of the obtained control algorithm for this model.

2. Statement of the problem

For description of nanosatellite orientation, we use two coordinate systems: body-fixed coordinate system $OX_bY_bZ_b$ and orbital coordinate system $OX_oY_oZ_o$. Axes OX_b, OY_b and OZ_b are main inertial axes of nanosatellite. Axis OZ_o connects center of Earth and nanosatellite mass center, axis OY_o is co-directed with orbital velocity, and axis OX_o complements the system to right-hand form.

After deployment on the orbit nanosatellites can acquire angular velocity up to few dozen degree fer second [19]. This leads to the problem of damping the angular velocities.

The control torque $\boldsymbol{M}$ in magnetic ADCS is the torque produced by a set of three magnetic coils interacting with the induction vector of the Earth's magnetic field $\boldsymbol{B}$:

$$\boldsymbol{M} = \boldsymbol{m} \times \boldsymbol{B}, \tag{1}$$

where $\boldsymbol{m}$ is the dipole moment of the coils in body coordinate system.

The classic B-dot algorithm forming the dipole moment $\boldsymbol{m}$ is calculated from the following equation [11-14]:

$$\boldsymbol{m} = -k\dot{\boldsymbol{B}}, \tag{2}$$

where k is the coefficient of proportionality, $\dot{\boldsymbol{B}}$ is the derivative of the geomagnetic vector in the body coordinate system.

After deployment and damping of the angular velocities, the ADCS faces the task of bringing the nanosatellite to a stable gravitational equilibrium position, with known requirements for orientation along the local vertical. But B-dot algorithm can damp only angular velocity only to values about orbital velocity.

In the current paper we propose two modifications of B-dot algorithm: the first one uses magnetometers and gyros as feedback, and the second one uses only magnetometers. The second option is reserved in case of gyro failure.

3. Modified B-dot algorithm

3.1. Modification of the B-dot algorithm using data from magnetometer and gyro sensor as feedback

For the first modification (*Modification 1*) of the B-dot algorithm, we use joint measurements of magnetometers and gyro sensors in the formation of feedbacks. With this approach, it is possible to use the relative velocity damping algorithm, which calculates the dipole moment as the vector product of the mismatch vector of the required and current angular velocities, and the geomagnetic vector:

$$\boldsymbol{m} = k\left[(\boldsymbol{\omega}_0 - \boldsymbol{\omega}) \times \boldsymbol{B}\right]. \tag{3}$$

During gravitational triaxial stabilization the nanosatellite moves relative to axis OY_o with orbital velocity ω_o, therefore the vector of the required angular velocity is:

$$\omega_{0,x} = 0,\ \omega_{0,y} = \omega_o,\ \omega_{0,z} = 0. \tag{4}$$

To achieve triaxial gravitational stabilization, the current should be supplied to a coil located along the axis with the second largest main moment of inertia OZ_b, since when the nanosatellite moves along a sun-synchronous orbit, the magnetic field lines form a small angle with nanosatellite velocity vector. Accordingly, the dipole moment has the following components:

$$m_x = 0,\ m_y = 0,\ m_z = -k\dot{B}. \tag{5}$$

The proposed modification consists of following mods:

(**Mode 1**) Damping of the initial angular velocities by B-dot algorithm;

(**Mode 2**) Keeping the value of the angular velocity close to the value of the orbital angular velocity using the relative velocity damping algorithm;

(**Mode 3**) Bringing nanosatellite to the required orientation by suppling current to the coil mounted on the axis with the second highest moment of inertia.

The condition for the transition from **Mode 1** to **Mode 2** is that the nanosatellite angular velocity module reaches the orbital angular velocity. The duration of **Mode 2** is chosen empirically and is six hours. The condition for disabling **Mode 3** is achievement of the required gravitational equilibrium position.

We derived the condition for disabling **Mode 3** under the assumptions of low angular velocity, in which the angular motion is divided into three independent channels [1]:

$$\begin{cases} \ddot{\beta} + \dfrac{3}{2}\dfrac{\mu}{R^3}\left(\dfrac{I_z - I_x}{I_y}\right)\sin 2\beta = 0 \\ \ddot{\psi} + 2\dfrac{\mu}{R^3}\left(\dfrac{I_y - I_x}{I_z}\right)\sin 2\psi = 0, \\ \ddot{\varphi} + \dfrac{1}{2}\dfrac{\mu}{R^3}\left(\dfrac{I_z - I_y}{I_x}\right)\sin 2\varphi = 0 \end{cases} \tag{6}$$

where μ is gravitational parameter of Earth, R is radius-vector of nanosatellite, I_x, I_y, I_z are main moments of inertia of nanosatellite, β is an angle between longitudinal axis and local vertical, ψ is a presession angle, φ is a proper rotation angle.

The integral of energy for system (6) is:

$$E = \frac{\dot{x}^2}{2} - \frac{1}{4}\boldsymbol{\Omega}^2 \cos 2\boldsymbol{x}. \tag{7}$$

where $\boldsymbol{x} = \beta, \psi, \varphi$, $\boldsymbol{\Omega} = \boldsymbol{\Omega}_\beta, \boldsymbol{\Omega}_\psi, \boldsymbol{\Omega}_\varphi$ is circular frequency, here:

$$\begin{cases} \Omega_\beta = \sqrt{3\mu\left(I_z - I_x\right)/(I_y R^3)} \\ \Omega_\psi = \sqrt{4\mu\left(I_y - I_x\right)/(I_z R^3)}. \\ \Omega_\varphi = \sqrt{\mu\left(I_z - I_y\right)/(I_x R^3)} \end{cases} \tag{8}$$

For stable equilibrium positions (0 and ±180 deg), equation (7) is written as:

$$E = \frac{\dot{x}_{\max}^{2}}{2} - \frac{1}{4}\boldsymbol{\Omega}^2, \tag{9}$$

and

$$E = -\frac{1}{4}\boldsymbol{\Omega}^2 \cos 2\boldsymbol{x}_{\max}. \tag{10}$$

The relationship between the maximum angles and maximum angular speeds goes from (9):

$$\begin{cases} \dot{\beta}_{\max} = \Omega_\beta \sin\beta_{\max} \\ \dot{\psi}_{\max} = \Omega_\psi \sin\psi_{\max} \\ \dot{\varphi}_{\max} = \Omega_\varphi \sin\varphi_{\max} \end{cases}. \tag{11}$$

The instantaneous angular velocity of the nanosatellite is the sum of the velocity components for each angle and the orbital angular velocity; therefore, the Euler equations are written as:

$$\begin{cases} \omega_x = \dot{\psi}\sin\beta + \dot{\varphi} + \omega_{orb}a_{12} \\ \omega_y = \dot{\psi}\sin\varphi\cos\beta - \dot{\beta}\cos\varphi + \omega_{orb}a_{22} \\ \omega_z = \dot{\psi}\cos\varphi\cos\beta + \dot{\beta}\sin\varphi + \omega_{orb}a_{32} \end{cases}, \tag{12}$$

where a_{12}, a_{22}, a_{32} are components of matrix A:

$$A = \begin{bmatrix} \cos\alpha & \sin\alpha\sin\psi & -\sin\alpha\cos\psi \\ \sin\alpha\sin\varphi & \cos\varphi\cos\psi - \cos\alpha\sin\varphi\sin\psi & \cos\varphi\sin\psi + \cos\alpha\sin\varphi\cos\psi \\ \sin\alpha\cos\varphi & -\sin\varphi\cos\psi - \cos\alpha\cos\varphi\sin\psi & -\sin\varphi\sin\psi + \cos\alpha\cos\varphi\cos\psi \end{bmatrix}. \tag{13}$$

In the approximation of small angles, relative to which the conclusion is drawn, we obtain the expression:

$$\begin{cases} \omega_{x,\max} = \Omega_2\psi_{\max}\beta_{\max} + \Omega_3\varphi_{\max} \pm \omega_o\psi_{\max} \\ \omega_{y,\max} = \Omega_2\psi_{\max}\varphi_{\max} - \Omega_1\beta_{\max} \pm \omega_o \\ \omega_{z,\max} = \Omega_2\psi_{\max} + \Omega_1\beta_{\max}\varphi_{\max} \pm \omega_o\varphi_{\max} \end{cases}. \tag{14}$$

Gravitational stabilization implies that the satellite will rotate in the orbital plane with an orbital angular velocity, therefore, we omit ω_o for $\omega_{y,\max}$ component and change signs from "-" to "+" to formalize lesser restrictions. Thus, the condition for the transition to a gravitational stable equilibrium position is achievement of following angular velocities relative to the ω_o:

$$\begin{cases} \omega_{x,\max} = \Omega_2\psi_{\max}\beta_{\max} + \Omega_3\varphi_{\max} + \omega_o\psi_{\max} \\ \omega_{y,\max} = \Omega_2\psi_{\max}\varphi_{\max} + \Omega_1\beta_{\max} \\ \omega_{z,\max} = \Omega_2\psi_{\max} + \Omega_1\beta_{\max}\varphi_{\max} + \omega_o\varphi_{\max} \end{cases}. \tag{15}$$

After reaching (15), the angular velocity components may at some moment cease to satisfy these constraints, then the algorithm will return to **Mode 3**.

3.2. Modification of the B-dot algorithm using magnetometer data as feedback

Another modification (*Modification 2*) of the algorithm is to use only magnetometers as feedback sensors. In this case the angular velocity module is estimated from the relative position of two measurements of the geomagnetic vector:

$$\left\|\boldsymbol{\omega}_k\right\| = \arccos\left(\frac{\boldsymbol{B}_{\mathrm{b},k-1}}{\left\|\boldsymbol{B}_{\mathrm{b},k-1}\right\|}\cdot\frac{\boldsymbol{B}_{\mathrm{b},k}}{\left\|\boldsymbol{B}_{\mathrm{b},k}\right\|}\right)\cdot\left(t_k - t_{k-1}\right)^{-1}. \tag{16}$$

The proposed modification consists of only two mods:

(**Mode 1**) Damping of the initial angular velocities by B-dot algorithm;

(**Mode 3**) Bringing nanosatellite to the required orientation by suppling current to the coil mounted on the axis with the second highest moment of inertia.

After the modulus of the angular velocity of the nanosatellite reaches the orbital velocity, **Mode 1** is switched on. We do not relate the angles and angular velocities near equilibrium position with magnetometers data to dissemble this mode, because at low angular velocities these data are strongly affected by orbital motion. We just let nanosatellite to damp to angular velocity as lower as possible. We empirically selected four days for duration of **Mode 3**.

4. Simulation of the proposed modifications of the B-dot algorithm

We provided investigation of proposed B-dot modifications on SamSat-ION nanosatellite. SamSat-ION is 3U CubeSat developing by Samara National Research University in Russia. The mission of SamSat-ION is dedicated to solving such problems as monitoring the parameters of the ionosphere by contact and remote methods. So, the scientific equipment of the satellite includes: a plasma parameter sensor [20] developed by the Institute of Applied Physics of the Russian Academy of Sciences, a navigation receiver for signals from satellite radio navigation systems, and an external triaxial magnetometer.

According to the requirements of the plasma parameters sensor, its plane must be perpendicular to the oncoming flow vector, for which three-axis gravitational stabilization is required (Fig. 1). When conducting experiments with a plasma parameter sensor, deviations of up to 30 deg from the local vertical and oscillations in the angle of proper rotation with an amplitude of up to 50 deg are permissible.

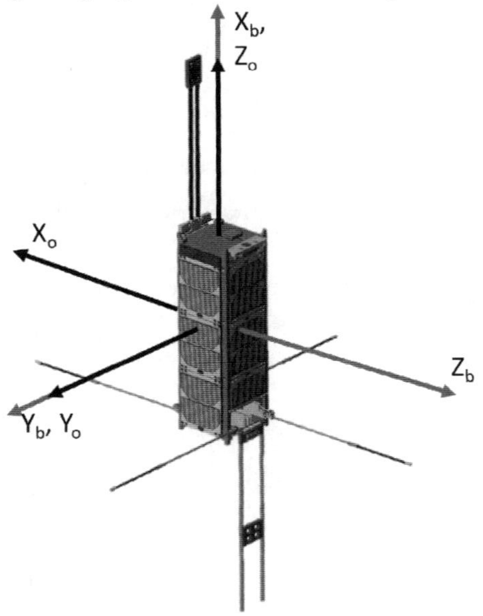

Figure 1. SamSat-ION in gravitational equilibrium position.

We provided statistical simulation includes 100 repetitions of the numerical integration of the equations of motion relative to nanosatellite mass center. The parameters are shown in table 1. After deployment the nanosatellite typically is spined along the transverse axes with a value of the about 1 deg/s [19]. The values of initial angular velocities are generated according to the normal distribution with mean M = 0 deg/s and variance σ2 = 1 (deg/s)2, and ωy > ωx > ωz due to properties of deployer. In simulation of Mode 1, Mode 2, and Mode 3 we used implementation of B-dot algorithm described in paper [21].

Table 1. Parameters of statistical simulation

Parameter	Value
Simulation period for *Modification 1*	6 days
Simulation period for *Modification 2*	10 days
Orbital altitude	550 km
Inclination	97 deg
Initial angle of precession	0 deg
Initial attack angle	90 deg
Initial proper rotation angle	0 deg
Modulus of initial angular velocity	1 deg/s
Damping coefficient in **Mode 1**	$3\cdot10^3$ $A\cdot m^2\cdot s/T$
Damping coefficient in **Mode 2**	$600\cdot10^3$ $A\cdot m^2\cdot s/T$
Damping coefficient in **Mode 3**	$1.5\cdot10^3$ $A\cdot m^2\cdot s/T$
Duration of **Mode 2** in *Modification 1*	6 hours
Duration of **Mode 3** in *Modification 2*	4 days
Maximum deflection angle β_{max}	30 deg
Maximum deflection angle ψ_{max}	30 deg
Maximum deflection angle φ_{max}	50 deg

4.1. Simulation results of modification of the B-dot algorithm using data from magnetometer and gyro sensor as feedback

Table 2 shows the distribution of the simulation results of *Modification 1* depending on the number of transitions from **Mode 3** to **Mode 4** and time required to accomplish the algorithm. There are only 11 cases when a stable equilibrium position was not achieved during the entire simulation period (the last column). The others 89 cases show stable average velocities $\tilde{\omega}_x$, $\tilde{\omega}_y$, and $\tilde{\omega}_z$, however the average angles $\tilde{\psi}$, $\tilde{\alpha}$, and $\tilde{\varphi}$ at the end of simulations show various character (table 3).

Table 2. Distribution of the resalts

Number of transitions from **Mode 3** to **Mode 4**	Total time of algorithm work, days						
	1	2	3	4	5	6	> 6
1	47	3	-	-	-	-	-
2	4	2	-	2	-	-	-
3	1	1	-	2	1	-	-
4	-	1	-	2	1	-	-
5	-	1	-	6	1	-	1
6	-	-	1	-	-	-	1
7	-	-	-	2	-	-	-
> 7	-	-	-	4	4	3	9

The first four lines of table 3 show that the nanosatellite achieved a stable equilibrium position, the angular velocity along OY_b axis close to orbital velocity ($\left|\tilde{\omega}_y\right| = \omega_o$). The fifth line shows the results where the nanosatellite did not achieve triaxial gravitational stabilization during the entire simulation period. The sixth line shows the cases, where the nanosatellite did not achieve the gravitational stabilization, but $\tilde{\omega}_y$ is stable and close to $\pm(\omega_o+\delta)$. This is a case of a quasi-plane motion. As a result, 73 of 100 results meet the requirements of gravitational stabilization and 27 not. The deviations from the local vertical do not exceed 10 deg and deviations in the angle of proper rotation do not exceed 40 deg. An example of one simulation is shown in Figure 2.

Table 3. The result of statistical simulation of *modification 1.*

№	$\tilde{\psi}$, deg	$\tilde{\alpha}$, deg	$\tilde{\varphi}$, deg	$\tilde{\omega}_x$, deg/s	$\tilde{\omega}_y$, deg/s	$\tilde{\omega}_z$, deg/s	Number
1	0	90	0	0	$+\omega_o$	0	24
2	0	90	180	0	$-\omega_o$	0	13
3	180	90	0	0	$-\omega_o$	0	14
4	180	90	180	0	$+\omega_o$	0	16
5	-	-	-	-	-	-	11
6	-	-	-	0	$\pm(\omega_o+\delta)$	0	22

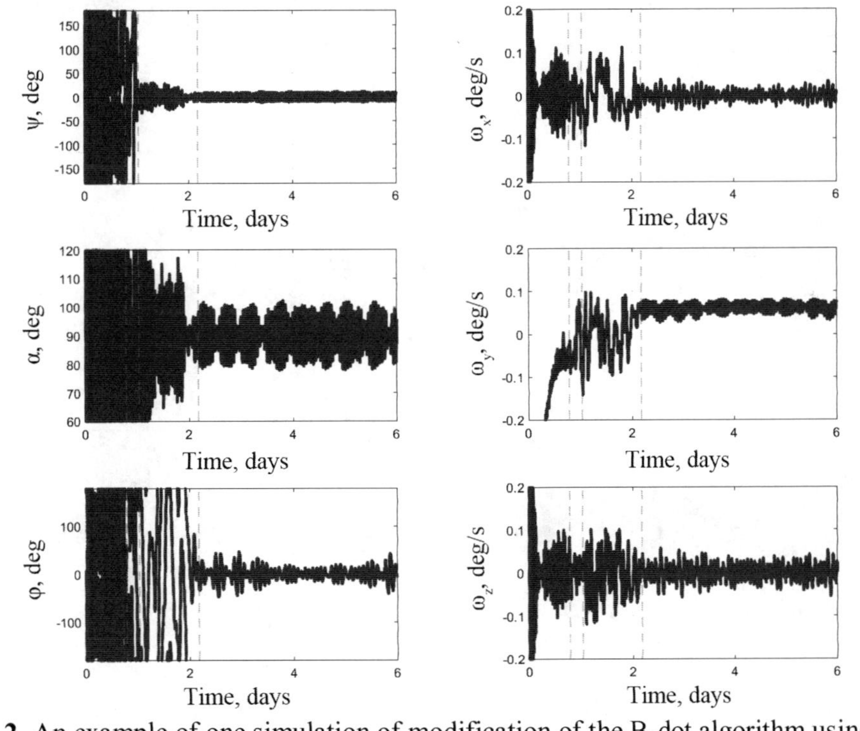

Figure 2. An example of one simulation of modification of the B-dot algorithm using data from magnetometer and gyro sensor as feedback.

4.2. Simulation results of modification of the B-dot algorithm using magnetometer data as feedback

The results of simulation of *Modification 2* are shown in table 4.

The first four lines of table 4 show that the nanosatellite achieved a stable equilibrium position, the angular velocity along OY_b axis close to orbital velocity ($\left|\tilde{\omega}_y\right| = \omega_o$). The fifth line represents the case of one-axis gravitational stabilization. The sixth line shows results where the nanosatellite did not achieve triaxial gravitational stabilization during the entire simulation period. The seventh line shows the cases of quasi-planar motion. As a result, 29 of 100 results met the requirements of gravitational stabilization, 17 of 100 results met more sparing requirement of one-axis stabilization and 54 did not meet. An example of one simulation is shown in Figure 3.

Table 4. The result of statistical simulation of *modification 2.*

№	$\tilde{\psi}$, deg	$\tilde{\alpha}$, deg	$\tilde{\varphi}$, deg	$\tilde{\omega}_x$, deg/s	$\tilde{\omega}_y$, deg/s	$\tilde{\omega}_z$, deg/s	Number
1	0	90	0	0	$+\omega_o$	0	9
2	0	90	180	0	$-\omega_o$	0	2
3	180	90	0	0	$-\omega_o$	0	11
4	180	90	180	0	$+\omega_o$	0	7
5	0	90	-	0	$\pm\omega_o$	0	17
6	-	-	-	-	-	-	3
7	-	-	-	0	$\pm(\omega_o+\delta)$	0	51

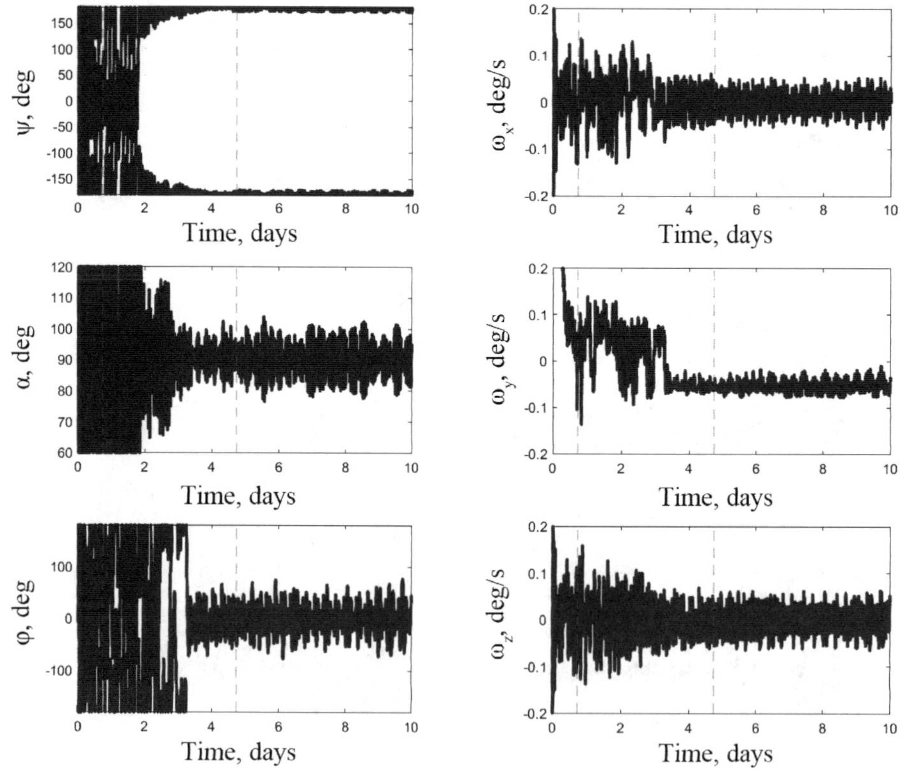

Figure 3. An example of one simulation of modification of the B-dot algorithm using magnetometer data as feedback.

5. Discussion and conclusion

The results of statistical simulation showed 67 successful cases of gravitational stabilization of 3U CubeSat out of 100 for *Modification 1* and 46 out of 100 for *Modification 2*. Also, in *Modification 2* simulation results there are 17 cases with one-axis gravitational stabilization, which in the case of less strict requirements can be considered a successful completion of the task. These results are actually is good, because classical B-dot algorithm is not intended to three-axis gravitational stabilization, only for damping initial angular velocities. Thereby, proposed modifications of B-dot algorithm show great possibility to provide three-axis gravitational stabilization of 3U CubeSat.

Nevertheless, it is important to increase the number of good cases of three-axis gravitational stabilization. There is an option how to it. It is possible to introduce of additional estimates of the nanosatellite attitude into the feedback of algorithm. So, having estimated the nanosatellite attitude, it is possible to understand whether it is in quasi-plane motion and whether it has one-axis gravitational stabilization that is, to determine which case is unsuccessful. If the case is unsuccessful, it is necessary to spin the nanosatellite and pass the circle of algorithm again. The number of unsuccessful cases will decrease with each pass of the algorithm.

Reference

[1] Belokonov I V and Timbai I A 2020 Dvizhenie nanosputnika otnositel'no centra mass na okolozemnyh orbitah: uchebnoe posobie (Samara, Samara university) 128 p. Russian.
[2] Desouky M A A and Abdelkhalik O O 2020 A new variant of the B-dot control for spacecraft magnetic detumbling *Acta Astronautica* **171** p. 14-22
[3] Bouwmeester J and Guo J 2010 Survey of worldwide pico- and nanosatellite missions, distributions and subsystem technology *Acta Astronautica* **67** pp. 854-862
[4] Silani E and Lovera M 2004 Magnetic spacecraft attitude control: a survey and some new results *Contr. Eng. Pract.* **13** pp. 357-371
[5] Wertz J R 1978 Spacecraft attitude determination and control *Springer* **7** p.9
[6] Sidi M J 1997 Spacecraft dynamics and control *Cambridge university press* **2**. p.7
[7] Desouky M A A, Prabhu K, Abdelkhalik O O 2018 On spacecraft magnetic attitude control. *Space flight mechanics meeting, AIAA* pp. 1-11
[8] Penkov V I and Roldugin D S 2018 Three-axis inertial magnetic orientation of satellite in gravity field *Keldysh Institute of Applied Mathematics preprints* **163** p.18
[9] Markley F L and Crassidis J L 2014 Fundamentals of spacecraft attitude determination and control *Space technology library Springer New York*
[10] Avanzini G, de Angelis E I, Giulietti F and Serrano N. 2019 Attitude control of Low Earth Orbit satellites byreaction wheels and magnetic torques *Acta Astronautica* **160** p. 625-634
[11] Ovchinnikov M U, Penkov V I, Roldugin D S and Ivanov D S 2016 Magnetic orientation systems of small satellites *Keldysh Institute of Applied Mathematics Moscow* 366 p
[12] Stickler A C and Alfriend K T 1976 Elementary magnetic attitude control system *Journal of Spacecraft and Rockets* **13** N.5 p. 282–287
[13] Stickler A C 1972 A Magnetic Control System for Attitude Acquisition *Ithaco, Inc.* Rep. N 90345
[14] Kramlikh A V, Melnik M E and Nikolaev P N 2016 Attitude determination and stabilization algorithms of the SamSat-218D nanosatellite *Proc. 23rd Saint Petersburg International Conference*
[15] Dechao R, Tao S, Lu C, Xiaoqian C and Yong Z 2013 Attitude control system design and on-orbit performance analysis of nano-satellite "Tian Tuo 1" *Chinese journal of aeronautics* **202**
[16] Zavoli A, Giulietti F, Avanzini G and de Matteris G 2016 Spacecraft dynamics under the action of Y-dot magnetic control law *Acta Astronautica* **122** p. 146-158
[17] Avanzini G, de Angelis E L and Giullietti F 2014 Spin-axis pointing of a magnetically actuated spacecraft *Acta Astronautica* **94** p. 493-501
[18] Cubas J and de Ruiter A 2020 Magnetic control without attitude determination for spinning spacecraft *Acta Astronautica* **169** p. 108-123
[19] Yudintsev V V 2015 Dynamics of CubeSat separation from a cubesat deployer *Polet* **8–9**. p. 10–15
[20] Galka A G, Yanin D V, Kostrov A V, Klimov S I and Novikov D I 2017 Resonant probe for measuring ionospheric plasma density and fluctuations on small satellites Section C. Dyagnosis of natural heterogeneous media and mathematical modeling BSFP-2017 p. 173–175
[21] Kramlikh A V, Lomaka I A and Nikolaev P N 2020 Damping control system design for SamSat nanosatellite platform *IOP Conference Series: Materials Science and Engineering* **862-2**

Acknowledgments
This work was supported by state funding allocated to the winners of the competition of scientific laboratories of higher educational institutions subordinate to the Ministry of Science and Higher Education of Russia (project No. 0777-2020-0018).

IOP Conf. Series: Materials Science and Engineering 1215 (2022) 012006 doi:10.1088/1757-899X/1215/1/012006

Test Results of the Long Baseline Navigation Solutions under a Large a Priori Position Uncertainty

V.V. Bogomolov

Concern CSRI Elektropribor, JSC, Russia

bogomolov.v.v@yandex.ru

Abstract. A method is proposed for long baseline navigation of autonomous underwater vehicles (AUV) to be used in the case of a large a priori position uncertainty. The new modified method is based on the iterated Kalman filter (IKF) working with different initial linearization points. The final solution is calculated by clustering and weighting the IKF results. This approach allows position estimates to be determined in accordance with the global maximum of posteriori probability density of coordinates. The test results obtained with the use of three beacons and an underwater vehicle are presented.

1. Introduction

Determination of coordinates by measuring ranges to sonar beacons (long baseline navigation) is one of the promising methods for AUV navigation, which is a special case of a more general method for determining the coordinates of a moving vehicle by measuring ranges to reference points [1–6]. Its difference from gravity-and-magnetic-field-aided navigation [7] and navigation using bathymetric data [8, 9] is that this method makes use of a limited amount of map information (coordinates of beacons). When AUV coordinates are determined using the ranges to the beacons, in the case of a large a priori uncertainty of coordinates, measurements have essential nonlinearity, which leads to a non-Gaussian posterior probability density of the vector being estimated, called, for brevity, a posteriori density. In this case, when the IKF is used, the estimate converges slowly, but the calculated and real accuracy characteristics do not coincide. The paper considers the solution of such a problem using the modified IKF. The main difference between the modified and traditional IKFs is the use of a set of initial linearization points instead of one initial point. The results of the modified and traditional IKFs are compared based on the test data.

2. Problem statement and its solution

For discrete point in time $i = 1, 2...$, we need to estimate the AUV's horizontal coordinates x_{1i}, x_{2i} in the local rectangular coordinate system by measuring the ranges to the beacons, which are formed using an a priori estimate of the speed of sound in water $\hat{c}_0$.

Let us introduce the state vector $X_i = [x_{1i}\ x_{2i}\ \ \Delta c]^{\mathrm{T}}$, where Δc is the zero-mean error in knowing the sound speed in water with root-mean-square error (RMSE) $\sigma_{\Delta c}$. The measurements are given as

$$y_i^j = T_i^j \hat{c}_0 = s_i^j(X_i) + \delta_i + v^j = \sqrt{(x_{1i}^j - x_{1i})^2 + (x_{2i}^j - x_{2i})^2 + (x_{3i}^j - x_{3i})^2} + \Delta c \bullet T_i^j + \delta_i + v_i^j,$$

Scientific adviser: D.A. Koshaev, Ph.D (Concern CSRI Elektropribor, JSC, Saint Petersburg)

where $j = 1..n_i$ is the number of the beacon; x_{1i}^j, x_{2i}^j are the known horizontal coordinates of the beacons; x_{3i}^j, x_{3i} are the known depths of the beacons and the AUV; T_i^j is the measurement of the signal travel time; δ_i and v_i^j are, respectively, the errors, common for all beacons, and uncorrelated for different beacons. δ_i and v_i^j are white noises with the RMSE σ_δ and σ_v, respectively.

The initial coordinates are given as

$$x_{10} = \hat{x}_{10}^0 + \Delta x_{10}, \quad x_{20} = \hat{x}_{20}^0 + \Delta x_{20},$$

where $\hat{x}_{10}^0$ and $\hat{x}_{20}^0$ are a priori estimates of the coordinates, and Δx_{10}, Δx_{20} are their zero-mean errors with the RMSE $\sigma_{\Delta x_{10}}$ and $\sigma_{\Delta x_{20}}$. Then the initial estimate and the covariance matrix have the form: $\hat{X}_0 = (\hat{x}_{10}^0\ \hat{x}_{20}^0\ 0)^{\mathrm{T}}$ and $P_0 = \begin{bmatrix} \sigma_{\Delta x_{10}}^2 & 0 & 0 \\ 0 & \sigma_{\Delta x_{20}}^2 & 0 \\ 0 & 0 & \sigma_{\Delta c}^2 \end{bmatrix}$.

It is assumed that the AUV has a dead reckoning system that uses measurements of velocity components V_1 and V_2 with zero-mean white noise errors w_1, w_2 with the RMSE σ_{w_1,w_1}. In reality, velocity components V_1 and V_2 are formed by recalculating the log velocity components using the heading. Note that heading error is not taken into consideration in this case and the equations of state vector dynamics have the form:

$$\dot{x}_1 = V_1 + w_1,$$
$$\dot{x}_2 = V_2 + w_2,$$
$$\Delta\dot{c} = 0.$$

3. Iterative Kalman filter

3.1. *Traditional IKF*

The following equations are used to solve the problem using the traditional IKF [10]:

$$P_i^k = ((P_{i/i-1})^{-1} + (H_i^k)^{\mathrm{T}} R_i^{-1} H_i^k)^{-1}, \; K_i^k = P_i^k (H_i^k)^{\mathrm{T}} R_i^{-1}, \; \hat{X}_i^k = \hat{X}_{i/i-1} + K_i^k (Y_i - S_i^k - H_i^k (\hat{X}_{i/i-1} - \hat{X}_i^{k-1})),$$

where $k = 1, 2...N$ is the iteration number, $\hat{X}_{i/i-1}$ is the estimate of the prediction and $P_{i/i-1}$ is the covariance matrix of the prediction errors, which are determined by the state vector dynamics equations and the results from the previous step $\hat{X}_{i-1}^N$ и P_{i-1}^N, $Y_i = (y_1 ... y_{n_i})^{\mathrm{T}}$, $S_i^k = (s_i^1(\hat{X}_i^{k-1}), ... s_i^{n_i}(\hat{X}_i^{k-1}))^{\mathrm{T}}$, $H_i^k = \left.\dfrac{ds_i(X_i)}{dX_i}\right|_{X_i = \hat{X}_i^{k-1}}$, $\hat{X}_i^0 = \hat{X}_{i/i-1}$, $R_i = \sigma_\delta^2 U + \sigma_v^2 I$ is the measurement noise covariance matrix, U is $n_i \times n_i$ matrix consisting of units, I is the $n_i \times n_i$ unit matrix.

3.2. *Description of the modified IKF*

The main difference between the modified IKF and the traditional one is the use of a set of initial linearization points instead of one initial point.

A prerequisite for obtaining a solution is the availability of measurements from at least three beacons.

Consider the modified algorithm in more detail. We pay special attention to the first moment in time and the subsequent period of time because the algorithm works differently on these two time intervals.

In the modified IKF, the following procedures are performed at the first step:

- a grid is formed from the linearization points in the vicinity of the a priori value of coordinates;
- for each point of the grid, an estimate is found by the traditional IKF;
- the total set of estimates is divided into disjoint clusters so that each cluster consists of close estimates, but the estimates between different clusters differ significantly;
- the centers of the clusters are determined;
- the cluster centers with the maximum weight (the value of posterior density) is determined by the way similar to the point mass method (grid method) [10]. Thus, the final estimate is expected to correspond to the global maximum of the posterior density.

At subsequent time points, it is not required to have measurements from at least three beacons, and the modified algorithm operates in the traditional IKF mode.

4. Test results

4.1. Initial data

Three AUV trajectories were used for tests (see Fig.1).

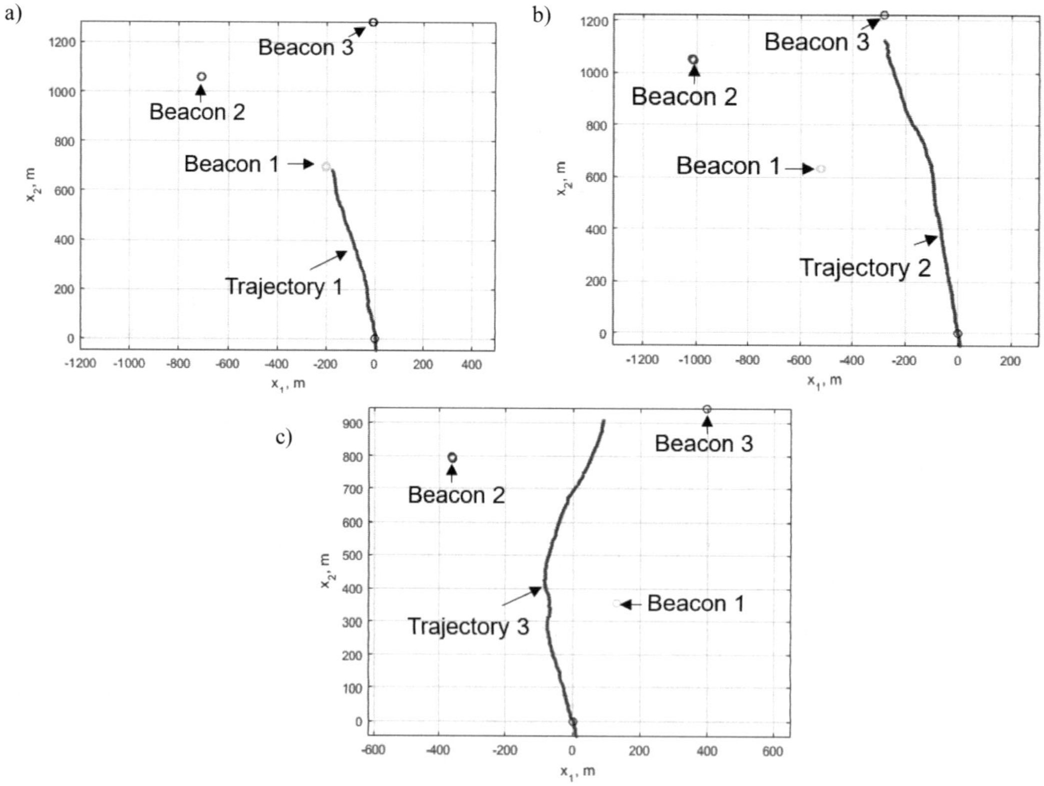

Figure 1. AUV trajectories: a) trajectory 1, b) trajectory 2, c) trajectory 3.

The values Δx_{10} and Δx_{20} are selected randomly. In this case, the RMSE take the following values: $\sigma_\delta = 7$ m, $\sigma_v = 20$ m, $\sigma_{\Delta c} = 10$ m/s, $\sigma_{\Delta x_{10}} = 500$ m, $\sigma_{\Delta x_{20}} = 500$ m, and $\sigma_{w_1, w_2} = 0.03$ m/s.

In the tests for each trajectories, we used the fixed values of the true coordinates of the AUV, measurements y_i^j, V_1 and V_2, as well as a priori estimate of the speed of sound in water $\hat{c}_0$, the value of T_i^j is determined as $T_i^j = y_i^j / \hat{c}_0$.

4.2. Results

The generalized characteristic of the modified IKF efficiency was obtained based on the simulations of $L = 1000$ random realizations of Δx_{10} and Δx_{20}. The number of iterations in the IKF was $N = 5$ in order to provide for the estimate convergence.

The results obtained with these methods were used to determine the real and calculated values of the RMSEs of coordinates at the first step for three trajectories (Fig. 2.) The calculated RMSE were obtained by averaging P_i^L for L realizations, and the real ones were calculated based on the errors of estimates $\hat{x}_1$ for L realizations.

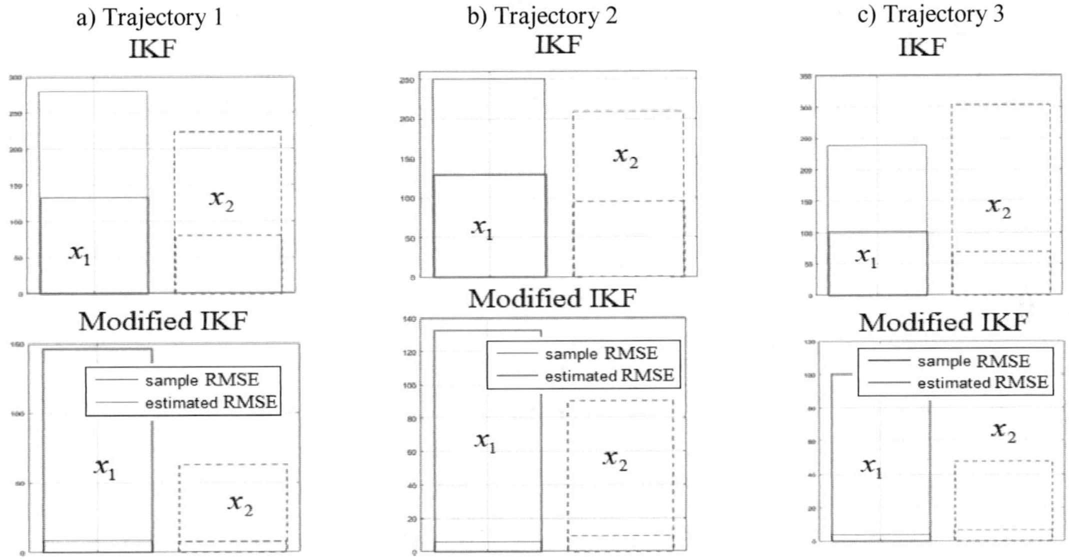

Figure 2. Real (---) and calculated (---) RMSE for three trajectories at the first point in time.

From the top diagrams in Fig. 2 it can be seen that in all cases, the real RMSE for the traditional IKF is several times higher than the calculated one, while in the modified IKF, the real error is much less than the calculated one, which suggests that the estimate of the traditional IKF does not always converge to the global maximum of posterior density. Note that it is missing in the case of the modified IKF. The calculated RMSE in the modified IKF is larger than the real one due to the fact that the experiment uses three realizations of measurements, which is insufficient for reliable statistics, and the error in estimating coordinates is mainly determined by the error in measurements, the realizations of which are fixed.

Note that the problems that arise in the traditional IKF at the first step remain for the rest of the time interval, while in the modified IKF, there were no problems at the first step and they do not arise later. This can be seen from Figs. 3–5, which show the calculated RMSE [9] and the real errors for the modified and traditional IKFs on all three trajectories.

Now, consider the efficiency of the algorithms in terms of their run time. The results for the algorithm runs for the first step are given in Table 1.

Table 1. Run time of the algorithms

IKF	Modified IKF
0.015 s	0.131 s where 0.126 s account for clustering

IOP Conf. Series: Materials Science and Engineering 1215 (2022) 012006 doi:10.1088/1757-899X/1215/1/012006

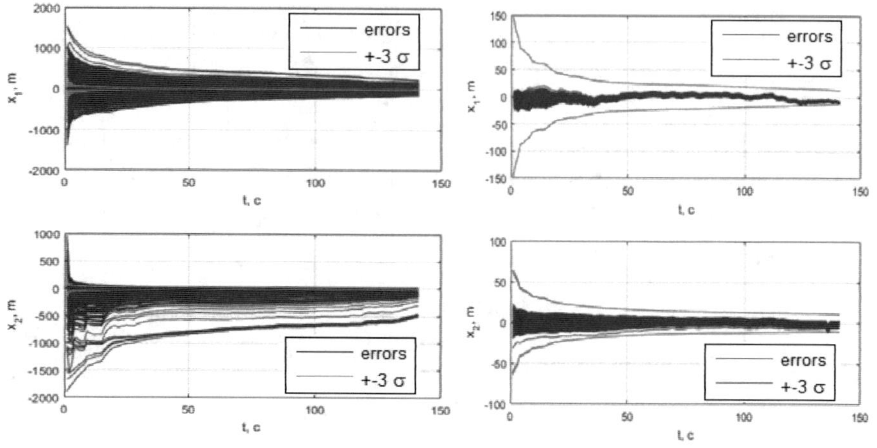

Figure 3. Calculated triple RMSE (---) and real errors (---) for the modified IKF (left) and the traditional one (right) for Trajectory 1.

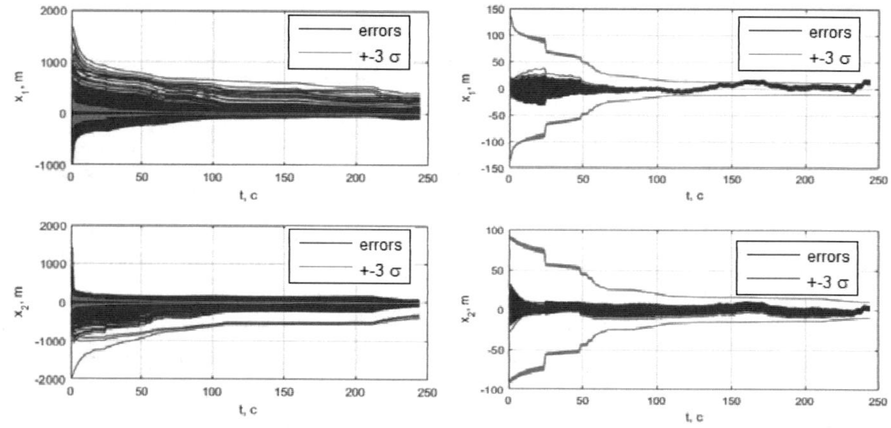

Figure 4. Calculated triple RMSE (---) and the real error (---) for the modified (left) and traditional IKF (right) for Trajectory 2.

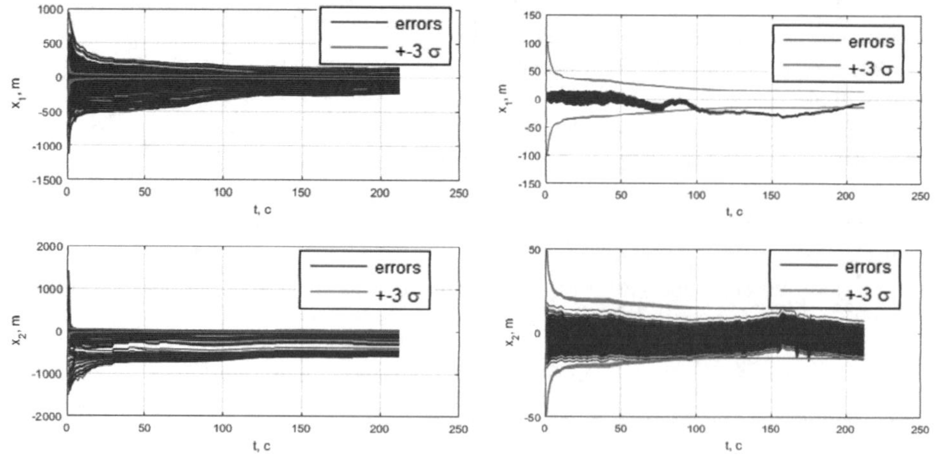

Figure 5. Calculated triple RMSE (---) and the real error (---) for the modified (left) and traditional IKF (right) for Trajectory 3.

The algorithm runtime was estimated using the standard Matlab software [11, 12]. It follows from the table that in the modified IKF, 90% of the time is spent on clustering. The clustering was performed using Matlab functions. We can significantly reduce the run time of the algorithm by improving the clustering procedure taken in to account the specifics of our purposes.

5. Conclusions

In the problem of long baseline navigation, by the example of three test trajectories with simulation of random a priori errors of coordinates, it is shown that the modified IKF has a significant advantage over the traditional IKF at the first and subsequent steps when there is a large a priori error of coordinates with the RMSE $\sigma_{\Delta x_{10}} = \sigma_{\Delta x_{20}} = 500$ m and at 1 km-ranges to the beacons. The calculated and real RMSE in the modified IKF are in good agreement. At the same time, the real RMSE is much higher than the calculated RMSE in the traditional IKF. The modified IKF is 10 times slower than the traditional one, which is due to the large consumption of computational resources for clustering.

References

[1] Paull, L., Saeedi, S., Seto M., Li H., AUV Navigation and Localization: A Review, 2014 *IEEE Journal of Oceanic Engineering*, **39**, no. 1, pp. 131–149.
[2] Kebkal,.K.G. and Mashoshin, A.I., AUV acoustic positioning methods, *Gyroscopy Navig.*, **8**, 80–89 (2017). https://doi.org/10.1134/S2075108717010059
[3] Matvienko, Yu.V., Inzartsev, A.V., Kiselev, L.V., and Shcherbatyuk, A.F., The prospects for improving efficiency autonomous underwater robots, Izvestiya YUFU. Tekhnicheskie nauki, 2016, pp. 123–141
[4] Stepanov, O.A., Estimation algorithms for navigation data processing. ITMO University, St.Petersburg, 2017.
[5] Koshaev, D.A., Multiple model algorithm for single-beacon navigation of autonomous underwater vehicle without its a priori position. Part 1. Mathematical formulation, *Gyroscopy Navig.* **11**, 230–243 (2020). https://doi.org/10.1134/S2075108720030037.
[6] Koshaev, D.A., Multiple model algorithm for single-beacon navigation of autonomous underwater vehicle without its a priori position. Part 2. Simulation, *Gyroscopy Navig.*, **4**, 319–332 (2020). https://doi.org/10.1134/S2075108720040069
[7] Stepanov, O.A. and Toropov, A.B., Nonlinear filtering for map-aided navigation part 2. Trends in the algorithm development, 2016 *Gyroscopy Navig.* **7**, 82–89.
[8] Dunaevskaya, K.V., Kiselev, L.V. and Kostousov, V.B. Study of a Method for Calculating the Current Accuracy in Map-Aided Navigation Problem. *Gyroscopy Navig.* **12**, 50–60 (2021). https://doi.org/10.1134/S2075108721010041
[9] Klyueva, S.F., Sintez algoritmov batimetricheskikh sistem navigatsii (Synthesis of Algorithms for Bathymetric Navigation Systems), Klyueva, S.F., Zav'yalov, V.V., Vladivostok: Morskoi Gos. Universitet, 2013. (in Russian) Stepanov, O.A., "Optimal and sub-optimal filtering in integrated navigation systems," in Aerospace Navigation Systems, A. Nebylov and J. Watson, Eds. New York, NY: John Wiley & Sons, Inc., 2016, pp. 392.
[10] Stepanov, O.A., Koshaev, D.A. A program for designing linear filtering algorithms for integrated navigation systems. IFAC Proceedings Volumes (IFAC-PapersOnline), 2011, 44 (1 PART 1), pp. 4256–4259
[11] https://www.mathworks.com

Slam Algorithms for Different Map Representations

R.U. Titov[1], A.V. Motorin[1,2]

[1] ITMO University, Saint-Petersburg, Russian Federation
[2] CSRI Elektopribor, Saint-Petersburg, Russian Federation

roman100895@rambler.ru

Abstract. The paper discusses the generalized simultaneous localization and mapping problem statement from the standpoint of the Bayesian approach and its relationship with algorithms for different map representations. The two-dimensional example describes the linearized simultaneous localization and mapping algorithm for the mobile platform in two-dimensional space.

1. Introduction

The main feature of simultaneous localization and mapping (SLAM), as the name of methods itself implies, is that the problems of positioning and map building are solved simultaneously. The solution to the SLAM problem is in demand for navigation of autonomous robots in GNSS denied environments such as closed rooms and under water [1–3].

One of the most common approaches to the solution of such problems is the stochastic Bayesian approach [1,4]. It involves the interpretation and finding the mathematical expectation and the covariance matrix for the conditional probability density function (PDF) of unknown values, namely, the object location and environment parameters. A set of environment parameters is usually called a map. In this case, the factorization of such a PDF and the direct calculation of parameters largely depend on a specific problem and, in particular, a map representation, which has given rise to many different algorithms. Depending on a map representation, SLAM is divided into feature-based SLAM, where a map is represented by a set of well-identifiable individual landmarks (features), and featureless SLAM, where individual landmarks are not distinguished.

The paper discusses the generalized SLAM problem statement from the standpoint of the Bayesian approach and its relationship with particular statements for different map representations. The two-dimensional example describes the linearized SLAM algorithm for the mobile platform in the 2D environment. In this case an environment is represented as a set of features and corresponds to the map defined by features coordinates. Possibilities for the concretization of the problems using other environments and map representations are discussed.

2. The general SLAM problem statement and solution

Consider an object moving in an environment, whose coordinates are supposed to be found. Consider an autonomous navigation system and sensors that provide additional measurements. The sensor measurements are related to a set of map parameters M, which describes the surrounding environment. Vector M is constant that considers the static environment. The solution to the SLAM problem is concluded in defining of object and map coordinates using a navigation system and sensors data.

Let $\mathbf{x}_i$ be a state vector or object coordinates at time i, where $i = 0..N$, $\mathbf{u}_i$ is the vector of input signals, $\mathbf{z}_i$ are the measurements that link to vector M defining an environment map. Let us introduce composite vectors of object coordinates $X_i = [\mathbf{x}_0,...,\mathbf{x}_i]$, input signals $U_i = [\mathbf{u}_0,..,\mathbf{u}_i]$ and measurements $Z_i = [\mathbf{z}_0,..,\mathbf{z}_i]$.

Consider the motion probability model is available in the form of the conditional PDF $f(\mathbf{x}_i / \mathbf{x}_{i-1},\mathbf{u}_i)$ and the prior PDF $f(\mathbf{x}_0)$ is known. The likelihood function $f(z_i / \mathbf{x}_i, M)$ is also considered to be known. In this case the SLAM problem may be represented as the generalized statement within Bayesian approach as the problem of searching the conditional (posteriori) PDF $f(\mathbf{x}_i, M / Z_i, U_i)$ by using given motion $f(\mathbf{x}_i / \mathbf{x}_{i-1},\mathbf{u}_i)$ and measurements $f(z_i / \mathbf{x}_i, M)$ models.

Using the well-known formulas, a posteriori PDF $f(\mathbf{x}_i, M / Z_i, U_i)$ may be represented as [5-7]:

$$f(\mathbf{x}_i, M / Z_i, U_i) = \eta f(z_i / \mathbf{x}_i, M) \int f(\mathbf{x}_i / \mathbf{x}_{i-1},\mathbf{u}_i) f(\mathbf{x}_{i-1}, M / Z_{i-1}, U_{i-1}) d\mathbf{x}_{i-1}, \quad (1)$$

where η is normalizing coefficient. Note that Eq. (1) is the Bayesian filter [3-5], which is also can be written as:

$$f(\mathbf{x}_i, M / Z_i, U_i) = \eta f(z_i / \mathbf{x}_i, M) f(\mathbf{x}_i, M / Z_{i-1}, U_{i-1}). \quad (2)$$

where

$$f(\mathbf{x}_i, M / Z_{i-1}, U_i) = \int f(\mathbf{x}_i / \mathbf{x}_{i-1},\mathbf{u}_i) f(\mathbf{x}_{i-1}, M / Z_{i-1}, U_{i-1}) d\mathbf{x}_{i-1} \quad (3)$$

and (3) corresponds the prediction PDF. Thus, formula (1) defines recurrence Bayesian filter for solving the SLAM problem.

3. Feature-based SLAM

Here is an example for the map represented as features in the two-dimensional (2D) environment with coordinates $M = \left[m_x^1, m_y^1, ..., m_x^L, m_y^L\right]^T$, where L is the number of features. Consider the navigation system on the object includes a speedometer and a compass and uses dead-reckoning method to recursively obtain the coordinates [8]. Let object carry a rangefinder that measures range and bearings to the features described by vector M.

Feature-based SLAM problem is formulated as a problem of object $\mathbf{x}_i = [x_i, y_i]^T$ and features coordinates estimation basing on navigation system data and rangefinder measurements, accumulated to time i.

True object coordinates achieved as the result of dead reckoning without usage of map data can be represented by equations:

$$\begin{cases} x_i = x_{i-1} + (\tilde{V}_i - \Delta V_i) \cdot \sin(\tilde{K}_i - \Delta K_i), \\ y_i = y_{i-1} + (\tilde{V}_i - \Delta V_i) \cdot \cos(\tilde{K}_i - \Delta K_i). \end{cases} \quad (4)$$

where $\tilde{K}_i$ is a heading measurement, $\tilde{V}_i$ is a speed measurement. The input signal vector is represented as $\mathbf{u}_i = \left[\tilde{K}_i, \tilde{V}_i\right]^T$. Coordinates calculated by (4) have an increase in errors in time due to speed and heading sensors errors. Dead reckoning system error may be calculated using on the linearized error model [8,10]:

$$\begin{cases} \Delta x_i = \Delta x_{i-1} + \tilde{V}_i \cdot \Delta t \cdot \cos \tilde{K}_i \cdot \Delta K_i + \Delta t \cdot \sin \tilde{K}_i \cdot \Delta V_i, \\ \Delta y_i = \Delta y_{i-1} - \tilde{V}_i \cdot \Delta t \cdot \sin \tilde{K}_i \cdot \Delta K_i + \Delta t \cdot \cos \tilde{K}_i \cdot \Delta V_i. \end{cases} \quad (5)$$

Presenting ΔV_i and ΔK_i as white Gaussian noise errors with covariances $\sigma^2_{\Delta K}, \sigma^2_{\Delta V}$, the prediction PDF may be written as:

$$f\left(\mathbf{x}_i / Z_{i-1}, U_i\right) = N\left(\begin{bmatrix} x_i \\ y_i \end{bmatrix}, \begin{bmatrix} \hat{x}_{i/i-1} \\ \hat{y}_{i/i-1} \end{bmatrix}, P^{\mathbf{x}}_{i/i-1}\right), \tag{6}$$

with the mathematical expectation and the covariance matrix are determined as

$$\hat{\mathbf{x}}_{i/i-1} = \hat{\mathbf{x}}_{i-1} + g\left(\tilde{K}_i, \tilde{V}_i\right), \tag{7}$$

$$P^{\mathbf{x}}_{i/i-1} = \Phi_i P^{\mathbf{x}}_{i-1} \Phi_i^T + \Gamma_i Q_i \Gamma_i^T, \tag{8}$$

where

$$\Phi_i = \begin{bmatrix} 1 & 0 \\ 0 & 1 \end{bmatrix},\ \Gamma_i = \begin{bmatrix} \tilde{V}_i \cos\tilde{K}_i & \sin\tilde{K}_i \\ -\tilde{V}_i \sin\tilde{K}_i & \cos\tilde{K}_i \end{bmatrix} \Delta t,\ Q_i = \begin{bmatrix} \sigma^2_{\Delta K} & 0 \\ 0 & \sigma^2_{\Delta V} \end{bmatrix},\ g\left(\tilde{K}_i, \tilde{V}_i\right) = \begin{bmatrix} \tilde{V}_i \Delta t \sin\tilde{K}_i \\ \tilde{V}_i \Delta t \cos\tilde{K}_i \end{bmatrix}.$$

It is worth noting that usage of a dead reckoning system is not the only possible here and instead of that an inertial navigation system may be used [9-12]. Here equations (4) are represented as the simplest example of the object coordinates calculation.

Since features coordinates are supposed to be constant, they are can be represented as unknown constant Gaussian random vectors. In this case the prediction of object coordinates and corresponding covariance matrices may be written as

$$\hat{M}_{i/i-1} = \hat{M}_{i-1},$$
$$P^M_{i/i-1} = P^M_{i-1}.$$

Thus, joint PDF (2) may be approximated by Gaussian in correspondence with the error model (5):

$$f\left(\mathbf{x}_i, M / Z_{i-1}, U_i\right) = N\left(\begin{bmatrix} \mathbf{x}_i \\ M \end{bmatrix}, \begin{bmatrix} \hat{\mathbf{x}}_{i/i-1} \\ \hat{M}_{i-1} \end{bmatrix}, P_{i/i-1}\right), \tag{9}$$

where

$$P_{i/i-1} = \begin{bmatrix} \Phi_i P^{\mathbf{x}}_{i-1} \Phi_i^T + \Gamma_i Q_i \Gamma_i^T & P^{\mathbf{x}_i M}_{i-1} \\ \left(P^{\mathbf{x}_i M}_{i-1}\right)^T & P^M_{i-1} \end{bmatrix}, \tag{10}$$

where $P^{\mathbf{x}_i M}_{i-1} = E\left\{\mathbf{x}_i M^T\right\}$ are correlation coefficients represented as zero $P^{\mathbf{x}_i M}_0 = 0$ at the initial time.

The likelihood function $f\left(\mathbf{z}_i / \mathbf{x}_i, M\right)$ is calculated basing on measurements $\mathbf{z}_i = [z_i^1, .., z_i^{2L}]^T$ to all L features for time i:

$$\begin{bmatrix} z_i^1 \\ \dots \\ z_i^L \\ z_i^{L+1} \\ \dots \\ z_i^{2L} \end{bmatrix} = \begin{bmatrix} \sqrt{(x_i - m_x^1)^2 + (y_i - m_y^1)^2} \\ \dots \\ \sqrt{(x_i - m_x^L)^2 + (y_i - m_y^L)^2} \\ \arctan\left(\dfrac{y_i - m_y^1}{x_i - m_x^1}\right) \\ \dots \\ \arctan\left(\dfrac{y_i - m_y^L}{x_i - m_x^L}\right) \end{bmatrix} + \begin{bmatrix} v_i^1 \\ \dots \\ v_i^L \\ v_i^{L+1} \\ \dots \\ v_i^{2L} \end{bmatrix}, \tag{11}$$

or in vector form

$$\mathbf{z}_i = s(\mathbf{x}_i, M) + \mathbf{v}_i,$$

where $\mathbf{v}_i = \left[v_i^1, ..., v_i^{2L} \right]^T$ is the vector of measurements white Gaussian noise. The Taylor series expansion of the likelihood functions leads to the Gaussian approximation of $f\left(\mathbf{z}_i / \mathbf{x}_i, M\right)$, whose parameters may be calculated using Extended Kalman Filter (EKF) [5-7,9,10]. In this case the measurements model is represented as:

$$\mathbf{z}_i \approx s(\mathbf{x}_i^o, M^o) + \left[\frac{\partial s(\mathbf{x}_i, M)}{\partial \mathbf{x}_i^T} \quad \frac{\partial s(\mathbf{x}_i, M)}{\partial M^T} \right]_{\substack{\mathbf{x}_i = \mathbf{x}_i^o \\ M = M^o}} \begin{pmatrix} \mathbf{x}_i - \mathbf{x}_i^o \\ M - M^o \end{pmatrix}. \tag{12}$$

Choose the prediction of dead reckoning system and features coordinates as the linearization point for $[\mathbf{x}_i^o, M^o]^T = [\hat{\mathbf{x}}_{i/i-1}, \hat{M}_{i-1}]^T$. This kind of a linearization lead to the calculation of residual measurements:

$$\tilde{\mathbf{z}}_i = \mathbf{z}_i - s(\hat{\mathbf{x}}_{i/i-1}, \hat{M}_{i-1}), \tag{13}$$

The Jacobi matrix $H_i = \left[\frac{\partial s(\mathrm{x}_i, M)}{\partial \mathrm{x}_i^T} \quad \frac{\partial s(\mathrm{x}_i, M)}{\partial M^T} \right]_{\substack{\mathrm{x}_i = \mathrm{x}_i^{Л} \\ M = M^{Л}}}$ is represented as:

$$H_i = \begin{bmatrix} \frac{\hat{x}_i - \hat{m}_{xi}^1}{\hat{d}_i^1} & \frac{\hat{y}_i - \hat{m}_{yi}^1}{\hat{d}_i^1} & \frac{-(\hat{x}_i - \hat{m}_{xi}^1)}{\hat{d}_i^1} & \frac{-(\hat{y}_i - \hat{m}_{yi}^1)}{\hat{d}_i^1} & 0 & \cdots & 0 \\ \cdots & \cdots & \cdots & \cdots & \cdots & \cdots & \cdots \\ \frac{\hat{x}_i - \hat{m}_{xi}^L}{\hat{d}_i^L} & \frac{\hat{y}_i - \hat{m}_{yi}^L}{\hat{d}_i^L} & 0 & 0 & \cdots & \frac{-(\hat{x}_i - \hat{m}_{xi}^L)}{\hat{d}_i^L} & \frac{-(\hat{y}_i - \hat{m}_{yi}^L)}{\hat{d}_i^L} \\ \frac{-(\hat{x}_i - \hat{m}_{xi}^1)}{(\hat{d}_i^1)^2} & \frac{-(\hat{y}_i - \hat{m}_{yi}^1)}{(\hat{d}_i^1)^2} & \frac{\hat{x}_i - \hat{m}_{xi}^1}{(\hat{d}_i^1)^2} & \frac{\hat{y}_i - \hat{m}_{yi}^1}{(\hat{d}_i^1)^2} & 0 & \cdots & 0 \\ \cdots & \cdots & \cdots & \cdots & \cdots & \cdots & \cdots \\ \frac{-(\hat{x}_i - \hat{m}_{xi}^L)}{(\hat{d}_i^L)^2} & \frac{-(\hat{y}_i - \hat{m}_{yi}^L)}{(\hat{d}_i^L)^2} & 0 & \cdots & 0 & \frac{\hat{x}_i - \hat{m}_{xi}^L}{(\hat{d}_i^L)^2} & \frac{\hat{y}_i - \hat{m}_{yi}^L}{(\hat{d}_i^L)^2} \end{bmatrix}, \tag{14}$$

where $\hat{d}_i^j = \sqrt{(\hat{x}_i - \hat{m}_{xi}^j)^2 + (\hat{y}_i - \hat{m}_{yi}^j)^2}$, $j = 1..L$.

Taking into account $f_v(v) = N(v; 0, R)$, the likelihood function may be represented as Gaussian:

$$f\left(\mathbf{z}_i / \mathbf{x}_i, M\right) \approx N\left(H_i \begin{bmatrix} \mathbf{x}_i \\ M \end{bmatrix}; \tilde{\mathbf{z}}_i + H_i \begin{bmatrix} \hat{\mathbf{x}}_{i/i-1} \\ \hat{M}_i \end{bmatrix}, R \right). \tag{15}$$

As the result of the Gaussian approximation for the prediction PDF (9) and likelihood function (15) the posterior PDF also may be approximated by Gaussian and its parameters may be calculated using EKF. As the result posterior PDF of the estimate (1) is represented as:

$$f\left(\mathbf{x}_i, M / Z_i, U_i\right) = N\left(\begin{bmatrix} \mathbf{x}_i \\ M \end{bmatrix}; \begin{bmatrix} \hat{\mathbf{x}}_i \\ \hat{M}_i \end{bmatrix}, P_i \right). \tag{16}$$

where

$$\begin{bmatrix} \hat{\mathbf{x}}_i \\ \hat{M}_i \end{bmatrix} = \begin{bmatrix} \hat{\mathbf{x}}_{i/i-1} \\ \hat{M}_{i/i-1} \end{bmatrix} + \mathbf{K}_i \left[\tilde{\mathbf{z}}_i \right], \tag{17}$$

$$\mathbf{K}_i = P_{i/i-1} H_i^T \left(H_i P_{i/i-1} H_i^T + R \right)^{-1}, \tag{18}$$

$$P_i = \left(E - \mathbf{K}_i H_i \right) P_{i/i-1}. \tag{19}$$

Thus, formulas (4), (10), (17) – (19) define EKF for the SLAM problem solution in the example. This method became the first variant of the SLAM problem solution and gained the currency in the beginning of 2000s. It has been referred as EKF-SLAM. The disadvantage of the method in particular is the quadratic growth of the computational complexity with the growth of features number because of the covariance matrix [4]. Here we withdraw the measurement data association from consideration. It can be also included in problem statement or solved separately [4]. Note that one-dimensional SLAM, considered in [13], does not require the approximation and may be solved using the linear Kalman Filter.

The SLAM methods were developed by the usage of PDF $f\left(X_i, M / Z_i, U_i\right)$ offered by [14] and advanced in [4]. Searching of $f\left(X_i, M / Z_i, U_i\right)$ seems intuitively more computationally complex, because whole object trajectory is taken into consideration. However, using the Bayes rule, it may be written:

$$f\left(X_i, M / Z_i, U_i\right) = f\left(X_i / Z_i, U_i\right) f\left(M / X_i, Z_i, U_i\right) = f\left(X_i / Z_i, U_i\right) f\left(M / X_i, Z_i\right). \tag{20}$$

For the map described by L independent features, as in example above, it may be rewritten as [1,4]:

$$f\left(X_i, M / Z_i, U_i\right) = f\left(X_i / Z_i, U_i\right) \prod_{l=1}^{L} f\left(m_x^l, m_y^l / X_i, Z_i\right). \tag{21}$$

The algorithms based on representation (21) are usually referred as Fast-SLAM [1,4]. The complexity of such algorithms increases logarithmically with the growth of features number [4,15]. Thus Fast-SLAM is more computationally effective than EKF-SLAM especially in the case of a large number of features. Fast-SLAM allows to use the occupancy grid map representation, where the dimension of map vector M is many times more than in map described by independent features. The map represented as occupancy grid can take only two values in any point, which is one if space is occupied and zero if it is free. Usually the whole map is divided in regular grid to specify the occupied cells. Thus the vector M can be represented a vector of random discrete independent variables [14]. The dimension of the vector corresponds to a number of grid cells. Finally this map representation allows rejecting the allocation of separated features and moving to so-called Featureless SLAM [2,3,16]. It also makes possible to consider the SLAM problem using geophysical fields such as magnetic or gravity anomalies in the similar way. This kind of fields map can be represented as a vector of random variables whose stochastic properties is defined by a known correlation function of a geophysical field [17]. Such a representation creates the preconditions for consideration within the framework of a unified formulation of the traditional problems of surveying geophysical fields and their use for navigation, which are currently considered as separate unrelated tasks [7,18,19].

4. Conclusions

The generalized SLAM problem statement from the Bayesian standpoint and its relationship with the SLAM algorithm for the mobile platform moving on a plane using the navigation dead reckoning system and rangefinder are shown. The EKF-SLAM algorithm is specified for the case using range and bearing measurements to the features, which coordinates compose the map of an environment. The possibility of generalizing the SLAM statement in case of another map representation is discussed.

References

[1] C. Cadena *et al.*, "Past, Present, and Future of Simultaneous Localization and Mapping: Toward the Robust-Perception Age," *IEEE Transactions on Robotics*, vol. 32, no. 6, pp. 1309–1332, Dec. 2016.

[2] N. Fairfield, G. Kantor, and D. Wettergreen, "Towards particle filter SLAM with three dimensional evidence grids in a flooded subterranean environment," in *Proceedings 2006 IEEE International Conference on Robotics and Automation, 2006. ICRA 2006.*, pp. 3575–3580.

[3] S. Barkby, S. Williams, O. Pizarro, and M. Jakuba, "An efficient approach to bathymetric SLAM," in *2009 IEEE/RSJ International Conference on Intelligent Robots and Systems*, 2009, pp. 219–224.

[4] M. Montemerlo, S. Thrun, D. Koller, and B. Wegbreit, "FastSLAM: A Factored Solution to the Simultaneous Localization and Mapping Problem," in *Eighteenth National Conference on Artificial Intelligence*, 2002, pp. 593–598.

[5] S. Sarkka, *Bayesian Filtering and Smoothing*. Cambridge University Press, 2013.

[6] J. V Candy, *Bayesian Signal Processing: Classical, Modern, and Particle Filtering Methods*. 2008.

[7] O. A. Stepanov and A. B. Toropov, "Nonlinear filtering for map-aided navigation. Part 1. An overview of algorithms," *Gyroscopy and Navigation*, vol. 6, no. 4, pp. 324–337, Oct. 2015.

[8] A. S. Nosov and A. V. Motorin, "Realizatsiya algoritmov oriyentatsii i navigatsii na chetyrekhkolesnoy mobil'noy platforme" (Implementation of orientation and navigation algorithms on a four-wheeled mobile platform), (in Russian). *Proc. of the XVII International Conference of Young Scientists, St. Petersburg: CSRI Elektropribor,* , 2015, pp. 212–220.

[9] O. A. Stepanov, "Optimal and sub-optimal filtering in integrated navigation systems," in *Aerospace Navigation Systems*, A. Nebylov and J. Watson, Eds. New York, NY: John Wiley & Sons, Inc., 2016, p. 392.

[10] P. D. Groves, *Principles of GNSS, Inertial, and Multisensor Integrated Navigation Systems.* Boston, London: Artech House, 2013, second edition.

[11] J. Crassidis and F. L. Markley, "Unscented Filtering for Spacecraft Attitude Estimation," *Journal of Guidance, Control, and Dynamics*, vol. 26, no. 4, pp. 536–542, 2003.

[12] V. N. Maksimov and A. I. Chernomorskii, "Integrated navigation and local mapping system for a single-axis wheeled module," *Gyroscopy and Navigation*, vol. 7, no. 3, pp. 285–295, Jul. 2016.

[13] A.V. Motorin, R.U. Titov, "Algoritmy odnovremennoj navigacii i kartografirovaniya dlya podvodnyh robotov" (Simultaneous navigation and mapping for undersea robots), (in Russian). *Proc. of the XXIII International Conference of Young Scientists, St. Petersburg: CSRI Elektropribor,* 2021. pp. 211-215.

[14] K. P. Murphy, "Bayesian Map Learning in Dynamic Environments," in *Proceedings of the 12th International Conference on Neural Information Processing Systems*, 1999, pp. 1015–1021.

[15] K. Murphy and S. Russell, "Rao-Blackwellised Particle Filtering for Dynamic Bayesian Networks," in *Sequential Monte Carlo Methods in Practice*, New York, NY: Springer New York, 2001, pp. 499–515.

[16] A. Eliazar and R. Parr, "DP-SLAM: Fast, Robust Simultaneous Localization and Mapping without Predetermined Landmarks," in *Proceedings of the 18th International Joint Conference on Artificial Intelligence*, 2003, pp. 1135–1142.

[17] M. Kok and A. Solin, "Scalable Magnetic Field SLAM in 3D Using Gaussian Process Maps," in *2018 21st International Conference on Information Fusion (FUSION)*, 2018, pp. 1353–1360.

[18] O. A. Stepanov and A. B. Toropov, "Nonlinear filtering for map-aided navigation Part 2. Trends in the algorithm development," *Gyroscopy and Navigation*, vol. 7, no. 1, pp. 82–89, Jan. 2016.

[19] V. G. Peshekhonov *et al.*, Sovremennyye metody i sredstva izmereniya parametrov gravitatsionnogo polya Zemli (Modern methods and equipment for Earth gravity field investigation) (in Russian). Saint-Petersburg, Russia: CSRI Elektropribor, 2017.

Acknowledgements

The work was financially supported by Russian Science Foundation project № 18-19-00627, https://rscf.ru/project/18-19-00627/.

IOP Conf. Series: Materials Science and Engineering 1215 (2022) 012008 doi:10.1088/1757-899X/1215/1/012008

Formation of an Additional Motion Control Loop for the Nanosatellite to Adapt its Onboard Model to the Current Operating Conditions

I. A. Lomaka
Samara University, Samara, Russia

igorlomaka63@gmail.com

Abstract. The paper presents the formation of additional feedback in the loop of the attitude control system of a nanosatellite. Feedback is based on the assessment of the inertial characteristics of the nanosatellite. The influence of the accuracy of knowledge of the inertial characteristics of a nanosatellite on the formation of an optimal control law in the problem of reorientation was estimated. Statistical modelling has been carried out to assess the effectiveness of nanosatellite on-board sensors in the problem of identifying the inertial characteristics of a nanosatellite. Recommendations for the selection of sensor`s characteristics and time interval of data collection have been formulated.

1. Introduction

Significant progress in microelectronics and minimization of spacecraft development cost led to the emergence of nanosatellites. Today, these spacecraft are increasingly used to implement a variety of space missions. The emphasis in the choice of a spacecraft class for solving modern problems of space exploration (global communication network [1], inspection of space debris objects [2-4], Earth remote sensing [5-7], etc.) is shifting towards the nanosatellites. The solution to these problems involves the creation of low-orbit nanosatellite`s constellations [8-11]. A nanosatellite-follower of the constellation should have increased manoeuvrability, power supply and autonomy.

Low energy and limited space for payload require various transformable structures on board the nanosatellite (deployable solar panels, retractable rods with equipment, etc.). In this case, the inertial and aerodynamic characteristics of the nanosatellite, as well as the margin of static stability, also noticeably change, which is important for low-orbit missions. In work [12] the influence of nanosatellite`s transformable structures on its aerodynamic characteristics was investigated. It is shown that neglect of transformable structures leads to an error in estimating the aerodynamic characteristics of nanosatellite in 15-20%.

The above circumstances require the consideration of the nanosatellite as an object with variable tensor of inertia, the position of the canter of mass and the centre of pressure. Changes in these parameters can affect the functioning of the nanosatellite`s attitude control system.

Thus, the implementation of modern space missions using the nanosatellite necessitates the solution on board the nanosatellite of a set of inverse parametric identification problems, the results of which can be used in the loop of the attitude control system to improve the quality of the control process.

This paper evaluates the efficiency of the additional feedback loop in the motion attitude system of the nanosatellite. This circuit is designed to assess the inertial parameters of the nanosatellite (Fig. 1).

IOP Conf. Series: Materials Science and Engineering 1215 (2022) 012008 doi:10.1088/1757-899X/1215/1/012008

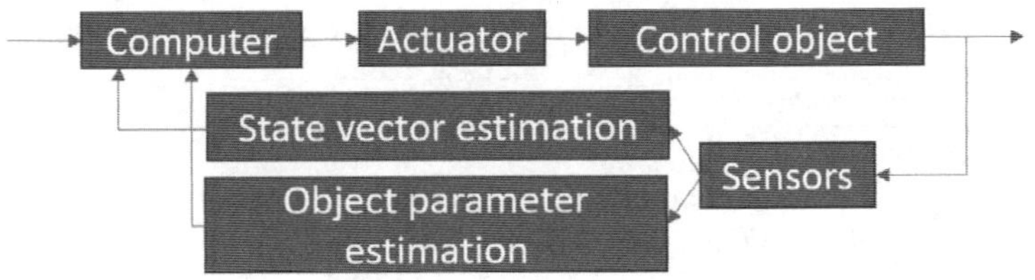

Figure 1. Structure of control system

2. Problem formulation

The angular motion of a nanosatellite in the plane case is described by the dynamic Euler equation of the following form:

$$\ddot{\alpha} = \frac{k_a}{I_t}\sin\alpha + \frac{k_g}{I_t}\sin 2\alpha + \frac{u(t)}{I_t} \tag{1}$$

where $k_a = C_x(\alpha)S_m Lq\Delta x$ - aerodynamic torque coefficient, $C_x(\alpha) = C_{x0}\left(|\cos\alpha| + \frac{4k}{\pi}|\sin\alpha|\right)$ - drag coefficient averaged over the angle of attack α, C_{x0} - drag coefficient, k - the ratio of nanosatellite lateral surface area to its representative area S_m, q - velocity head, Δx - static stability margin, L - nanosatellite length, $k_g = \frac{3\mu^2}{2\boldsymbol{r}^6}(I_t - I_l)$ - gravitational torque coefficient, I_t - transverse moment of inertia, I_l - longitudinal moment of inertia, μ - Earth gravitation parameter, $\boldsymbol{r}$ - nanosatellite radius-vector, $u(t)$ - control torque.

The goal is to find such a nominal trajectory and the corresponding control torque, which will ensure the transfer of nanosatellite from the initial position to the final one within a certain time interval T and will satisfy the chosen criterion of optimality.

The problem of reorientation of a nanosatellite was considered for the following required boundary conditions:

$$\begin{cases} \alpha(0) = 180°, \dot{\alpha}(0) = 2°/\text{s}, \\ \alpha_r(T) = 0°, \dot{\alpha}_r(T) = 0°/\text{s}. \end{cases}$$

where T – time of reorientation. $\alpha(T)_r, \dot{\alpha}(T)_r$ - angle and angular velocity that must be realized at the end of reorientation process.

The nanosatellite turn time will be taken as a quarter turn. For an orbit with an altitude of 415 km the span of a quarter turn takes T = 1400 s. At this time, the process reorientation can be considered fast relative to the period of nanosatellite motion along orbit, so we can neglect the decrease in flight altitude due to actions of the disturbing forces. This means that the coefficient of gravitational torque will remain constant in time, and the coefficient the aerodynamic torque will depend only on the current angle of attack of the nanosatellite.

We will choose a CubeSat-3U nanosatellite as the spacecraft under study. On control is imposed a limitation due to the maximum torque that can be created by the actuators of the nanosatellite. Specifications of the nanosatellite are shown in Table 1.

The control torque is written in the form of a Fourier series:

$$u(t) = A_0 + A_1\cos\left(\frac{2\pi}{2T}t + \varphi_1\right) + A_2\cos\left(\frac{4\pi}{2T}t + \varphi_2\right) \tag{2}$$

Table 1. Weight, size and inertia characteristics of nanosatellite

Characteristics	Value
Mass, kg	2
Transverse moment of inertia, kgm^2	0.01975
Longitudinal moment of inertia, , kgm^2	0.004
Length, m	0.3
Midsection area, m^2	0.01
Static stability margin	0.06
Maximum control torque, Nm	$5 \cdot 10^{-6}$
Mass, kg	2
Transverse moment of inertia, kgm^2	0.01975
Longitudinal moment of inertia, kgm^2	0.004
Length, m	0.3

Coefficients $A_0, A_1, A_2, \varphi_1, \varphi_2$ were determined as a result of solving the problem of finding the minimum of a functional of the form:

$$J = v_\alpha \left(\alpha(T)_r - \alpha(T)_o\right)^2 + v_{\dot\alpha}\left(\dot\alpha(T)_r - \dot\alpha(T)_o\right)^2 + v_u \int_0^T u^2(t)\mathrm{dt} \to \min, \tag{3}$$

where $v_\alpha = (1°)^{-2}, v_{\dot\alpha} = (1°/s)^{-2}, v_u = (1Nm^2 \cdot s)^{-1}$ - weight coefficients, $\alpha(T)_o, \dot\alpha(T)_o$ - obtained finale angle and angular velocity. Minimization (3) was carried out by the method of differential evolution [13].

The measurement system, which includes a combination of a triaxial angular velocity sensor and a triaxial magnetometer, is widely used on nanosatellites. This is due to the fact that these measuring instruments have a low cost, acceptable accuracy and small dimensions.

The measurements obtained from triaxial angular velocity sensor $\boldsymbol{\omega}(t)$ are simulated by solving equation (1). Stochastic measurement errors (measurement noise) are simulated by adding $\boldsymbol{w}_\omega(t)$ - white noise with root mean square σ_ω. Thus, the mathematical model of measurements for the angular velocity sensor will be written in the form:

$$\boldsymbol{C}_\omega(\boldsymbol{b},t) = \boldsymbol{\omega}_{\text{meas}}(t) = \boldsymbol{\omega}(\boldsymbol{b},t) + \boldsymbol{w}_\omega(t). \tag{4}$$

In the case of using a triaxial magnetometer, the measurement model looks like:

$$\boldsymbol{C}_B(\boldsymbol{b},t) = \boldsymbol{B}_{\text{meas}}(t) = A(\boldsymbol{b},t)\boldsymbol{B}_{orb} + \boldsymbol{w}_B(t), \tag{5}$$

where $\boldsymbol{w}_B(t)$ - white noise with root mean square σ_B, in general $\boldsymbol{b}$ is a vector of estimated parameters but in this paper $\boldsymbol{b} = I_t$ is a scalar value, $A(\boldsymbol{b},t)$ – transition matrix from orbital to body coordinate system [15]; $\boldsymbol{B}_{orb}$ – vector of the Earth's magnetic field induction in the orbital coordinate system, calculated according to IGRF model.

Additional feedback in the motion control system of the nanosatellite is based on the estimation of the vector of the nanosatellite parameters $\boldsymbol{b}$. The problem of identifying nanosatellite vector $\boldsymbol{b}$ is formulated as follows: to evaluate the vector using a set of measurement vectors $\boldsymbol{C}(\boldsymbol{b},t)$ carried out on board the NS. The problem posed is solved by minimizing the following objective function:

$$J(\boldsymbol{b},T) = \sum_{i=1}^{N}\left[\left(\boldsymbol{C}(t_i) - \hat{\boldsymbol{C}}(\boldsymbol{b},t_i)\right)^T \left(\boldsymbol{C}(t_i) - \hat{\boldsymbol{C}}(\boldsymbol{b},t_i)\right)\right] \to \min, \tag{6}$$

where $\boldsymbol{C}(t_i)$ – measurements vector, in this paper $\boldsymbol{C}(t_i)$ is $\boldsymbol{C}_\omega(\boldsymbol{b},t_i)$ or $\boldsymbol{C}_B(\boldsymbol{b},t_i)$, their joint use is not intended, $\hat{\boldsymbol{C}}(\boldsymbol{b},t_i)$ – simulated measurements vector, $N = \frac{T}{\Delta t}$ – number of measurements; $\Delta t = 10s$ – discreteness of measurements, T – time interval of data collection.

To estimate the $\boldsymbol{b}$ vector identification error, the following expression is used [14]:

$$\frac{\sigma_{meas}^2}{\delta_b} = \frac{\partial J(\boldsymbol{b},T)}{\partial \boldsymbol{b}},$$

$$\delta_b = \left(\frac{\partial J(\boldsymbol{b},T)}{\partial \boldsymbol{b}}\right)^{-1} \sigma_{meas}^2, \qquad (7)$$

where σ_{meas} is equal to σ_ω or σ_B (depends on sensor type), δ_b is equal to δI in this paper.

3. Simulation results

As a result of solving the reorientation problem, the nominal trajectory of the nanosatellite and the corresponding control law were obtained. Similar problems were solved in the case of a 10% error in knowledge of the moment of inertia I_t. The obtained control laws were used to reorient the nanosatellite with a nominal value of the moment of inertia I_t. The phase trajectories of the nanosatellite are shown in the figure 2.

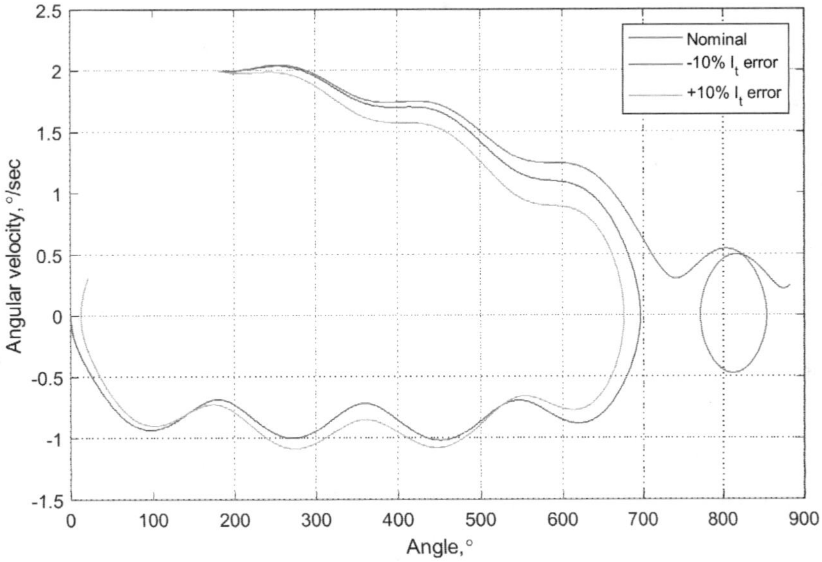

Figure 2. Phase portraits for different control laws

It is shown that an estimation error of I_t 10% leads to significant errors in the control of the nanosatellite. This demonstrates the importance of using additional feedback in the control loop.

Figures 3 and 4 show the results of statistical modeling of additional feedback on the estimate of the transverse moment of inertia of the nanosatellite. Simulation was carried out for an angular velocity sensor and a magnetometer. The initial angular velocity of the nanosatellite was taken to be equal to the orbital angular velocity for a given altitude.

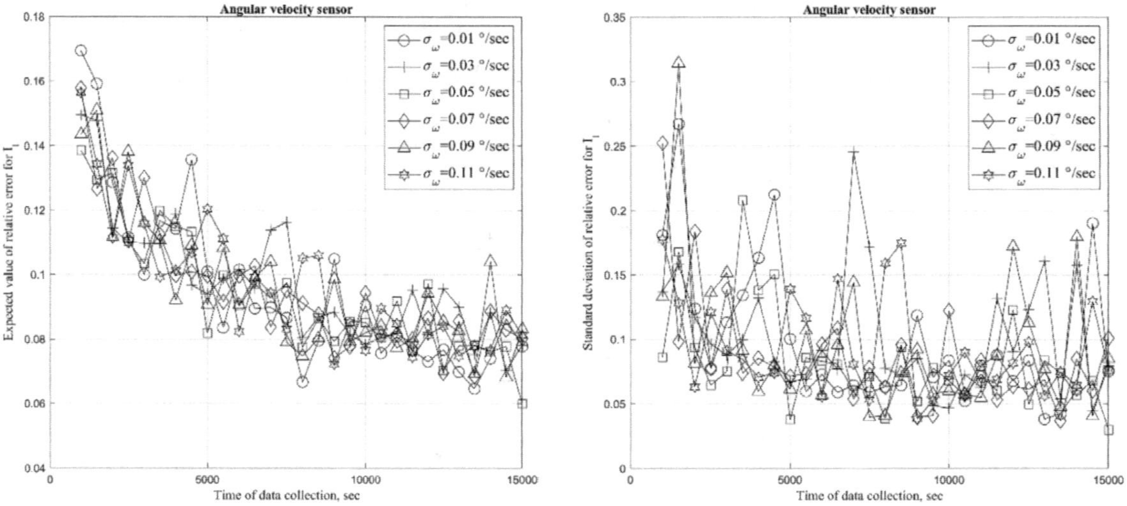

Figure 3. Simulation results for angular velocity sensor (relative error $\delta I_t / I_t$)

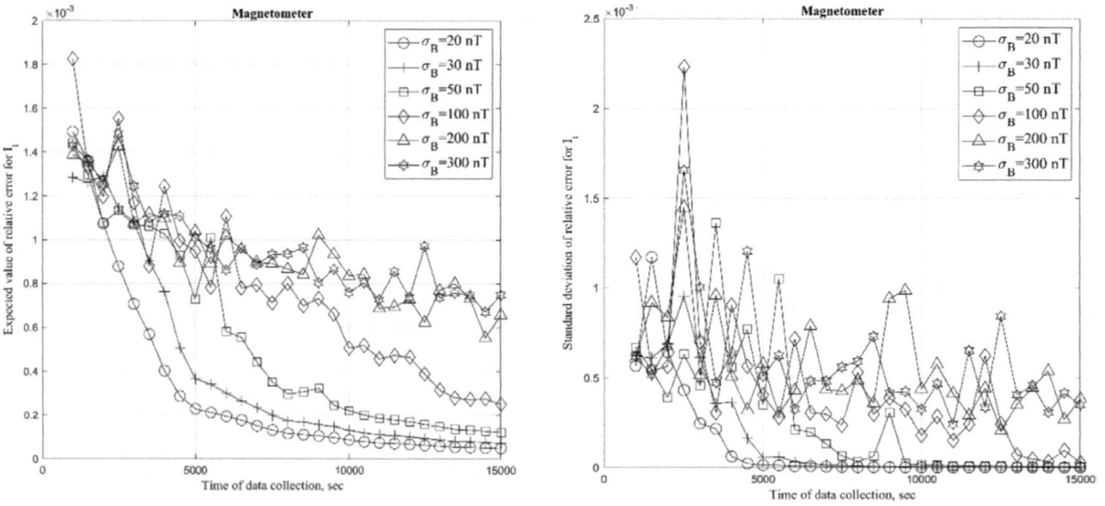

Figure 4. Simulation results for angular velocity sensor (relative error $\delta I_t / I_t$)

The results obtained showed the inefficiency of using the angular velocity sensor, which is associated with the high relative measurement error of this sensor. The best results are shown by a magnetometer with a measurement error of 20-100 nT. With the accumulation of a sample of measurements over a time interval of the order of three turns (15000 s), it is possible to obtain an estimate of the transverse moment of inertia with a relative error of less than 5%. Such error is sufficient for the implementation of modern missions using nanosatellites

4. Conclusion

The use of additional feedback in the loop of the nanosatellite control system, which monitors the parameters of the nanosatellite, has been substantiated. It is shown that the presence of this feedback makes it possible to improve the quality of the motion control of the nanosatellite. Evaluation of the effectiveness of measuring instruments allows us to formulate a recommendation on the use of a magnetometer for evaluating the parameters of a nanosatellite, as well as to select an interval for collecting measuring information of the order of three orbital turns.

References

[1] Arifin J 2021 Study of CUBESAT systems for IoT *12th Int. Renewable Engineering Conf.* (Amman, Jordan) pp 1–3

[2] Spiller D, Magionami E, Schiattarella V, Curti F, Facchinetti C, Ansalone L and Tuozzi A 2020 On-orbit recognition of resident space objects by using star trackers *Acta Astr.* vol 177 pp 478 – 496

[3] Riano-Rios C, Bevilacqua R and Dixon W 2021 Adaptive control for differential drag-based rendezvous maneuvers with an unknown target *Acta Astr.* vol 181 pp 733–740

[4] Han N, Luo J, Ma W and Yuan J 2018 Integrated identification and control for nanosatellites reclaiming failed satellite, *Acta Astr.* vol 146 pp 387–398

[5] Farrag A, Othman S, Mahmoud T Y and ELRaffiei AY 2021 Satellite swarm survey and new conceptual design for Earth observation applications *The Egyptian J. of Remote Sens. and Space Sci* vol 24 pp 47–54

[6] Santilli G, Vendittozzi C, Cappelletti C, Battistini S and Gessini P 2018 CubeSat constellations for disaster management in remote areas, *Acta Astr.* vol 145 pp 11–17

[7] Guerra A, Ferreira A S, Costa M, Nodar-López D and Agelet F A 2018 Integrating small satellite communication in an autonomous vehicle network: A case for oceanography, *Acta Astr.* vol 145 pp 229–237

[8] Lippe C and D'Amico S 2021 Optimal spacecraft swarm reconfiguration through chief orbit refinement, *Acta Astr.* vol 183 pp 162–175

[9] Koenig A W and D'Amico S 2021 Robust and Safe N-Spacecraft Swarming in Perturbed Near-Circular Orbits, *J. of Guidance, Contr. and Dyn.* vol 41 pp 1643–62

[10] Koenig A W and D'Amico S 2018 Safe spacecraft swarm deployment and acquisition in perturbed near-circular orbits subject to operational constraints, *Acta Astr.* vol 153 pp 297–310

[11] Lim Y 2020 In-Plane Formation Reconfiguration with Radial Maneuvers, *J. of Guidance, Contr. and Dyn.* vol 43 pp 1881–92

[12] Elisov N A, Barinova E V and Lomaka I A 2021 Improved estimation of aerodynamic characteristics of a complex geometry nanosatellite *Eng. J: Sci. and Inn.* vol 118 pp 1–17

[13] Elisov N A, Ishkov S A, Lomaka I A and Shakhov V G 2020 Influence of non-equilibrium reactions on the optimization of aerothrust aeroassisted maneuver with orbital change, *Chinese J. of Aeronautics* vol 33 pp 2133–45

[14] Belokonov I V and Lomaka I A 2020 Methodology of parametric identification of nanosatellite angular motion model, *Kosmonavtika i raketostroenie* vol 6 pp 134–145

Acknowledgments

Thanks to Nikolay Elisov for his help in modelling the procedure for forming the optimal control law of a nanosatellite in the reorientation problem.
The reported study was funded by RFBR, project number 20-08-00617A.

IOP Conf. Series: Materials Science and Engineering 1215 (2022) 012009 doi:10.1088/1757-899X/1215/1/012009

Model of Signal Mark Detection in a Passive Sonar System

V.V. Prokopovich, A.V. Shafranyuk

Concern CSRI Elektropribor, JSC, Saint Petersburg, Russia

wm.prokopovich@yandex.ru, andrey.shafranyuk@mail.ru

Abstract. Modeling of broadband and narrowband signal mark detection is widely used for sonar and radar systems. In this work, the problem is considered in relation to hydroacoustics. The paper describes the formation of a stream of correctly detected and false signal marks and calculation of estimates of their parameters, taking into account the antenna characteristics as well as the processing parameters of the system being simulated. Also considered are the realistic distribution of false signal marks by heading angles and the influence of the Doppler effect on the estimation of the mark parameters. The resulting model can be used in simulation systems, in the formation of a stream of detected signal marks, and the development of tracking algorithms. The model can be also used for predictive calculations that determine the probability of detecting signal sources and their characteristics

1. Introduction

Detection of signal marks is an integral part of data processing in hydroacoustic systems (HAS). A signal mark is a set of detected signal parameters, including such coordinates as heading and elevation angles, and their root-mean-square errors (RMSE), as well as the signal-to-noise (S/N) ratio, signal level, and signal response width. Detection of a signal and, accordingly, estimation of its parameters are carried out when the S/N ratio exceeds a certain threshold. As a result, a signal mark is formed [1].

Since in practice it is difficult to obtain a sufficient amount of real data from a hydroacoustic system in different conditions, it is done by simulation. Simulation of signal detection and estimates of their parameters is used to solve various problems, but, first of all, to develop and test algorithms for estimation of motion tracks of the objects being observed and to display primary data in software of hydroacoustic systems [2, 3].

The approach described in this paper makes it possible to calculate the parameters of the signal mark, i.e., to form its model both for a mark corresponding to a correctly detected signal and for a false mark (in relation to their heading angles). Besides the basic parameters, the calculation procedure for which was known earlier, the proposed model of signal mark detection allows taking into account a number of additional factors, namely: distribution of heading angles of false marks and accounting for narrowband signals in broadband direction finding, which provides a more adequate stream of false marks (including those from narrowband signals) in debugging the tracking analysis and adjustment of appropriate parameters.

2. General requirements for the signal mark detection. Problem statement.

The set of input data for the signal mark formation model is defined as follows:

- motion parameters of the HAS carrier and the object being observed:
  - speed of the carrier and the object being observed $V\left(i_{crr}\right), V\left(i_{obj}\right)$;

- their headings: $\gamma(i_{crr})$, $\gamma(i_{obj})$;
- the heading angle to the object being observed – $\alpha_{obj}(i_{obj})$ and a set of elevation angles to the multipath noise signal from the target – $\{\beta_{obj}(i_{obj}, i_{beam})\}$;

- noise signal parameters of the object being observed at the HAS carrier location:
  - power of the continuous signal spectrum in the HAS frequency bands (FB): $P_{sgnl.FB}(i_{FB})$;
  - power of discrete samples (DS) of the signal spectrum, their frequencies and widths: $P_{sgnl.DS}(i_{DS})$, $f_{sgnl.DS}(i_{DS})$ and $df_{sgnl.DS}(i_{DS})$;
  - the distributed interference power in the FB $P_{clutter.FB}(i_{FB})$ and power density spectrum $P_{clutter}(f)$;
- parameters of the HAS antenna array:
  - geometry of the antenna array. In the case that the basic shapes of antenna arrays are flat and linear and close to them, the antenna geometry is described by its width and height: L_{ant}, H_{ant};
  - geometric mean frequencies of the FBs and their boundaries: $f_{FB.avg}(i_{FB})$, $f_{FB.min}(i_{FB})$, $f_{FB.max}(i_{FB})$;
  - sampling frequency of the signal f_{samp};
- parameters of signal processing in the HAS:
  - period of spectral analysis (SA): T_{SA};
  - the number of SA accumulations: $N_{SA.acc}$;
  - false alarm probability (false signal detection – marks): $P_{dtctd.false}$.

Simulation of these parameters was aimed to form signal marks which, depending on the conditions, correspond to

- broadband signal:

$$M_{BBF}(i) = \{I, P, \alpha, d\alpha, \beta, d\beta, A\}^T, \tag{1}$$

where I is the S/N ratio in the frequency band; P is the signal power estimate in the frequency band,; α is the bearing estimate; $d\alpha$ is the RMSE of the bearing estimate, β is the elevation estimate,; $d\beta$ is the RMSE of the elevation estimate; A is the signal response width;

- narrowband signal:

$$M_{NBF}(i) = \{I, P, \alpha, d\alpha, \beta, d\beta, f, df, A\},^T \tag{2}$$

where f is the estimate of the central frequency; df is the RMSE of the central frequency estimate.

In addition, the model is supposed to provide for the possibility of generating false marks. It is also necessary to ensure that the model can function in real time.

Variation of input parameters allows signal marks to be generated for different types of HAS in different situations and underwater conditions, which makes the model universal and provides for a full development-and-test cycle of tracking algorithms.

In what follows, we give the dependencies between the input and output (1–2) data of the signal mark detection model used to form a stream of correct and false marks. Note that we do not consider the calculation of false mark parameters in this paper; however, it is planned to be done in the future.

3. Model of signal mark detection in broadband direction-finding mode

By correct marks, we mean the marks corresponding to the signals from noisy objects arriving at the antenna array from different directions (according to the beam structure of sound propagation in the environment). In other words, potentially, it is possible to form a mark for each object for each of the beams in each of the frequency bands of a hydroacoustic system, wherein broadband detection (by the total broadband energy) of signals takes place. In turn, the beams—the paths of sound waves propagation in the refracting medium—are determined by the model of this medium, for which the BELLHOP model [4] is most suitable to consider. Note that further, all calculations are given for each correct mark on the beam and in the corresponding FB; in so doing, indices i_{obj}, i_{beam} and i_{FB}, as well as i_{DS}, are omitted.

For the given conditions the fact of signal mark detection is simulated by the Monte Carlo method. In this case, the probability of detection depends on the S/N ratio and the probability of false alarm. It is defined by the formula:

$$P_{dtctd.true} = F_{norm}(I - F_{norm}^{-1}(1 - P_{dtctd.false})); \tag{3}$$

where $Pdtctd.true$ is the probability of detection; F_{norm} is the normal distribution function; F^{-1}_{norm} is the inverse normal distribution function, and I is the S/N ratio of the signal.

In the model, the S/N ratio for a broadband signal is simulated as a random process with a normal distribution. Their mathematical expectation and RMSE are calculated as follows:

$$\begin{gathered} I = norm(I_{FB}, \sigma_I) \\ I_{FB} = \sqrt{\frac{P_{\text{sgnl.FB}}}{P_{\text{clutter.FB}}} \cdot T_{SA} N_{SA.acc}}, \sigma_I = \frac{\overline{I}}{6} \\ P_{\text{clutter.FB}} = \int_{f_{FB.\min}}^{f_{FB.\max}} \frac{P_{clutter}(f)}{K_{conc}(f, \alpha_{obj}, \beta_{obj})} \cdot df \end{gathered} \tag{4}$$

where $\overline{I}$ is the expectation of the S/N ratio; $norm(x,y)$ is the normally distributed random number with expectation x, and RMSE y; σ_I is the RMSE of the S/N ratio estimate; $K_{conc}(f, \alpha_{obj}, \beta_{obj})$ is the antenna array concentration factor. Moreover, for example, for a flat antenna,

$$K_{conc}(f, \alpha_{obj}, \beta_{obj}) = 10^{-6} \frac{4\pi \cdot L_{ant} H_{ant}}{c^2} \cdot f^2.$$

The broadband signal power is calculated as follows:

$$P_{FB} = norm(P_{signal.FB}, \sigma_{P.FB}), \sigma_{P.FB} = \frac{P_{signal.FB}}{6}, \tag{5}$$

where $\sigma_{P.FB}$ is the expectation and the RMSE of the broadband noise power estimate.

The estimates of the bearing and elevation angle are simulated in a similar way, which is why we will only consider estimation of the bearing. It is also simulated as a random process with a normal distribution. The RMSE is calculated using the transformed formula (3.3.44) from [5, p. 290] as follows:

$$\alpha = norm\left(\alpha_{obj}, \sigma_{\alpha.obj}\right),$$

$$\begin{gathered} \sigma_\alpha = \sqrt{\frac{1}{2N_{SA.acc} T_{SA}} \int_{f_{FB.\min}}^{f_{FB.max}} h_{DC.hor}\left(\alpha_{obj}, \beta_{obj}, f\right) \frac{1 + I_{SNR}(f)}{I_{SNR}^2(f)} df} \approx \\ \approx \sqrt{\frac{1}{2N_{SA.acc} T_{SA}} h_{DC.hor}\left(\alpha_{obj}, \beta_{obj}, f_{FB.avg}\right) \frac{1 + I_{FB}}{I_{FB}^2}} \end{gathered} \tag{6}$$

where $\bar{\alpha}$, σ_α are the expectation and the RMSE of the bearing estimate; I_{SNR} is the S/N ratio related frequency f.

The signal response width of the mark is estimated as the width of the 0.5-level response from the S/N ratio. The shape of the hydroacoustic system response to broadband noise can be approximated using the Gaussian function (Fig. 1):

$$I(\alpha) = I_{FB} \cdot exp\left(-\frac{\alpha^2}{\left(K_{norm} \cdot h_{DC.hor}\left(f_{FB.avg}, \alpha_{obj}, \beta_{obj}\right)\right)^2} \right), \tag{7}$$

where $f_{FB.avg}$ is the center frequency; $K_{norm} = 1.45$ is the normalizing ratio.

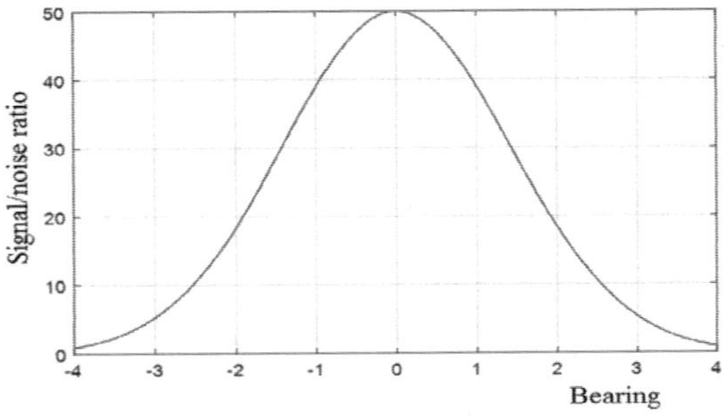

Figure 1. Sonar system response to broadband noise.

Using the formula for the full width at half maximum of the Gaussian function, we calculate the signal response width as follows:

$$A = 2 \cdot \sqrt{2 \cdot ln\,2} \cdot K_{norm} \cdot h_{DC.hor}(f_{avg}, \alpha_{obj}, \beta_{obj})\,. \tag{8}$$

The above formulas are the basic formula for the signal mark detection model. However, it is necessary to take into account a number of specific effects. One of them, namely, accounting for narrowband signals in the broadband direction-finding channel, is considered below.

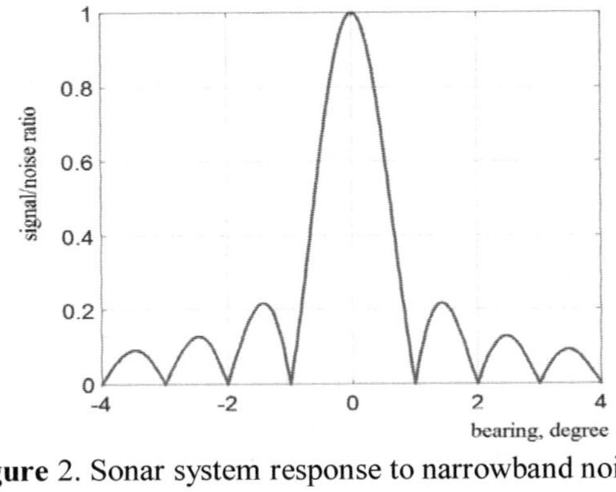

Figure 2. Sonar system response to narrowband noise.

The main aim of broadband noise direction finding is to detect broadband noise signals, but if a narrowband process has a sufficient power, it can be detected in a broadband [1]. To estimate the parameters of such marks, we use the sinc function to approximate the shape of the HAS response to a narrowband signal in Fig. 2:

$$I(\alpha) = I_{FB.NB} \cdot \mathrm{sinc}\left(\frac{\pi \cdot \alpha}{h_{DC.hor}(\mathrm{f}_{ss}, \alpha_{obj}, \beta_{rcv}) \cdot K_{norm}} \right), \tag{9}$$

where $I_{FB.NB}$ is the S/N ratio of the narrowband signal in the frequency band, calculated by the formula similarly to formula (4):

$$I_{FB.NB} = \sqrt{\frac{P_{\mathrm{sgnl.DS}}}{P_{\mathrm{clutter.FB}}} \frac{df_{signal.DS}}{f_{FB.\max} - f_{FB.\min}} \cdot T_{SA} N_{SA.acc}} .$$

In this case, the signal power is calculated similarly to formula (5), where the formula $\frac{df_{signal.DS}}{f_{FB.\max} - f_{FB.\min}} P_{\mathrm{sgnl.DS}}$ must be used instead of $P_{\mathrm{sgnl.FB}}$. In turn, the probability of detecting such a mark is estimated by formula (3), and the other parameters are similar to fornulas (6)–(9).

Now, consider the distribution of false marks in the frequency band. Signal marks are called false if the noise power exceeds the detection threshold specified by the Neumann-Pearson criterion in accordance with $P_{dtctd.false}$. The number of such marks in a processing cycle is inextricably linked with the probability of a false alarm:

$$N_{false.FB} = 4 P_{dtctd.false} f_{FB.\mathrm{avg}} f_{samp} \frac{L_{ant} H_{ant}}{c^2}, \tag{10}$$

where c is the speed of sound.

Thus, on average, the number of false marks can be determined by (10).

The traditional approach to modeling of false signal marks assumes an even distribution of marks in bearing and elevation throughout the HAS [6] entire field of view. However, due to the heterogeneity of the antenna array, its noise immunity significantly depends on the direction of signal reception (Fig. 3).

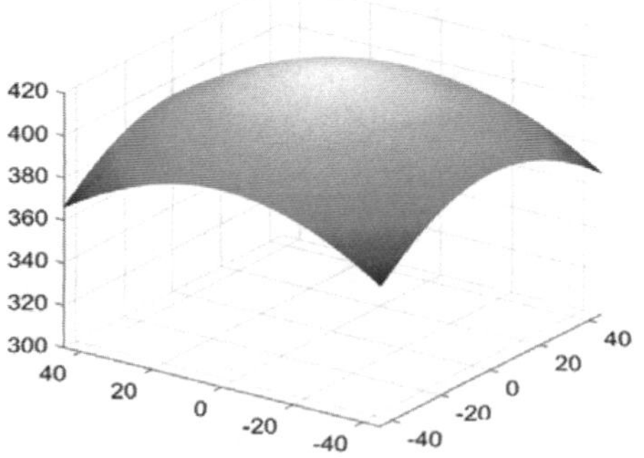

Figure 3. Distribution of the concentration factor.

In this case, the model under consideration will have a significantly greater consistency when the distribution density of false marks is inversely proportional to the concentration factor.

$$F_{pdf.mark.false}(\alpha, \beta) \approx K_{conc}^{-1}(\alpha, \beta). \tag{11}$$

4. The proposed model of signal mark detection in narrowband direction-finding mode

Modeling of signal mark detection in the narrowband noise direction-finding mode is carried out similarly to that in the broadband noise direction-finding mode using formulas (3)–(10). Nevertheless, in the case of narrowband direction-finding, marks have an additional parameter – frequency f. The frequency estimate is modeled as a random realization of a random number with a normal distribution. Then, taking into account the Doppler effect, the expectation and RMSE can be written as follows:

$$f = norm(f_{DS} \cdot k_{dop}, \sigma_f) \\ \sigma_f = max(\Delta f_{res}, d f_{DS}) \tag{12}$$

In this case, the Doppler coefficient is calculated by the well-known full formula (relative to the environment):

$$k_{dop} = \left(1 - \frac{V_{rcv.rad} \cdot cos(\beta_{rcv})}{c_{rcv}}\right)\left(1 - \frac{V_{src.rad} \cdot cos(\beta_{src})}{c_{src}}\right), \tag{13}$$

where $V_{rcv.rad} = \left(\{\cos(\alpha);\sin(\alpha)\}, V(i_{crr})\{\cos(\gamma(i_{crr}));\sin(\gamma(i_{crr}))\}\right)$ is the radial speed of the HAS carrier, and $V_{src.rad} = -\left(\{\cos(\alpha);\sin(\alpha)\}, V(i_{obj})\{\cos(\gamma(i_{obj}));\sin(\gamma(i_{obj}))\}\right)$ is the radial speed of the object being observed; β_{rcv} and β_{src} are the elevation angles for the beam along which the signal propagates, respectively, at the HAS carrier location and at the location of the object being observed, and c_{rcv} and c_{src} are the corresponding speeds of sound.

5. Conclusions

The paper describes the formation of a model for the functioning of primary data processing in a hydroacoustic noise direction-finding station. The proposed approach is also applicable to the development of similar models in radars. The model, together with all the necessary data, makes it easy to simulate a stream of signal marks (including false ones) in order to debug tracking analysis algorithms or solve problems of primary data display in sonar and radar stations.

The proposed approach to the model formation is universal in the sense that it involves a wide range of variable parameters considered in the first section.

However, a fairly wide range of issues has remained beyond the scope of this publication, such as accounting for the mutual influence of the detected signals and, accordingly, the parameters of their marks, especially in the case of a close angular arrangement; description of the distribution density function of a stream of false marks; estimation of the consistency of both the model as a whole and its individual components (contribution of one or another part of the model to its overall consistency).

The authors are planning to consider these issues in the works which are to be published in the nearest future.

References

[1] H. Peyvandi, M. Farrokhrooz, H. Roufarshbaf, and S.-J. Park. SONAR Systems and Underwater Signal Processing: Classic and Modern Approaches. *Sonar Systems* 2011, pp. 173–206

[2] Voronina N.G. and Shafranyuk A.V. Algorithm for constructing trajectories of maneuvering object based on bearing-only information using the Basis Pursuit method // Journal of Physics: Conference Series. 13th Multiconference on Control Problems (MCCP 2020). 2021.

[3] Bykova, V.S., Mashoshin, A.I. & Pashkevich, I.V. Safe Navigation Algorithm for Autonomous Underwater Vehicles. *Gyroscopy Navig.* 12, 86–95 (2021).

[4] Porter M.B., Bucker H.P. Gaussian beam tracing for computing ocean acoustic fields // J. Acoust. SWoc. Am., 1987. V.82 №4, P. 1249-1359.

[5] Malyshkin G.S. Optimal'nye i adaptivnye metody obrabotki gidroakusticheskih signalov [Optimal and adaptive methods for processing hydroacoustic signals]; OAO "Koncern "CNII "Elektropribor". St-Peterburg :, 2009

[6] Etter, Paul C., Underwater acoustic modeling and simulation. // Boca Raton: CRC Press, Taylor & Francis Group, 2018 154-175

Acknowledgments

The reported study was funded by RFBR, project number 19-08-00253.

IOP Conf. Series: Materials Science and Engineering 1215 (2022) 012010 doi:10.1088/1757-899X/1215/1/012010

Design of the Capacitive Dual-Mass Micromechanical Gyroscope

A.V. Styazhkina, A.A. Belogurov, Ya.V. Belyaev, A.T. Tulaev

Concern CSRI Elektropribor, JSC, Saint Petersburg, Russia

anna_yakimov@mail.ru

Abstract. Development of micromechanical inertial sensors have made it possible to use them in the navigation and motion control systems. This application area imposes strict requirements on sensors. One of the ways to meet the requirements and to improve the gyroscope characteristics is to apply a dual- or multi-mass architecture of a gyroscope sensing element. This paper presents the results of dual-mass micromechanical gyroscope with a measurement range of ±450°/s design. The complex design method, including simulation at the system level, model refinement based on the results of finite element modelling, and modelling of individual electronic blocks at the circuit level, is described.

1. Introduction

Due to small size and low power consumption, as well as low cost in mass production, micromechanical gyroscope (MMG) is used to determine angular velocity as an integral part of modern compact navigation and motion control systems. This application area imposes strict requirements on sensors [1–2]:

- wide measurement range, more than ± 200°/s;
- low non-linearity, less than 0.01 %;
- wide dynamic range, more than 100 dB;
- bias repeatability (in condition of temperature variations from minus 60 to plus 85 °C, mechanical shock and vibration, etc.), less than ±100°/h.

The main parts of capacitive MMG are the following:

- sensing element (SE), destined to convert an input angular velocity to a capacitance change;
- application-specific integrated circuit (ASIC), destined to convert a capacitance change to a digital data and to control the SE in case of close-loop sensor.

To meet the requirements it is necessary to apple solutions at the architectural level of the SE, as well as the ASIC:

- using of multi-mass sensing element [3–6, 8–9];
- using of specific integrated circuits, in particular close-loop circuits with a delta-sigma modulator in the control loop [7], including frequency tuning [8] and quadrature compensation control loops [9];
- using temperature stabilization systems, etc.

This paper presents the results of dual-mass MMG with a measurement range of ±450°/s design, produced with cavity-SOI technology. The complex design method, including simulation at the system level, model refinement based on the results of finite element modelling, and modelling of individual electronic blocks at the circuit level, is described.

2. MMG design method

Complication of sensor architecture, impossibility to prototype, necessity to joint model ASIC blocks with a SE, necessity to consider influence of external influence factors on characteristics, as well as influence of technological parameters require to design micromechanical sensors based on the accurate mathematical model. The existing design methods and mathematical models of inertial micromechanical sensors [10–14] have a number of drawbacks:

- ASIC blocks do not consider or consider in a simplified way;
- SE is represented as a number of lumped parameters (stiffness of elastic suspension, capacitance of electrode structure), determined analytically or by finite element modeling;
- models do not consider experimental data.

In this regard, the method proposed in this work is based on system model of sensor (Fig. 1) and on complexing it with results of sensor components detail modeling (finite element modeling of sensing element, schematic modeling for IC blocks), test results and technological imperfections.

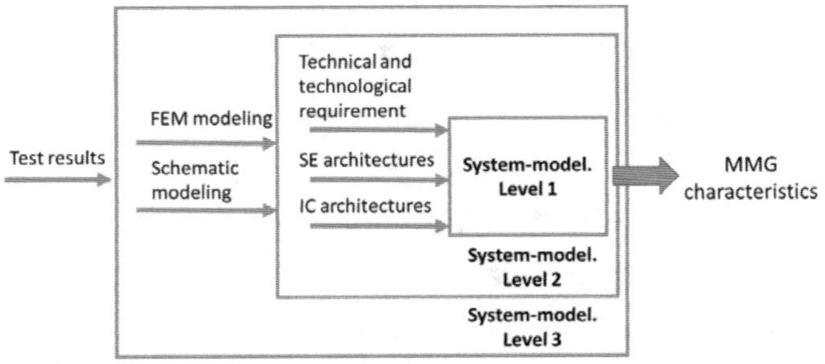

Figure 1. Structural scheme of system model complexing in design process

Thus, a prototype of a digital twin sensor is implemented in the design process. This approach allows estimating MMG output characteristics and to optimize parameters of SE and IC in considering with output characteristics. Detailed description of IC design based on proposed method is given in [15], main points of SE design process are shown in this paper.

3. System model

On the first level system model is formed by following blocks:

- transfer function of SE mechanical part;
- coefficients describing SE transducer;
- transfer function of main IC blocks.

As an example, Fig.2 shows the part of first level system model that describes MMG secondary oscillation loop.

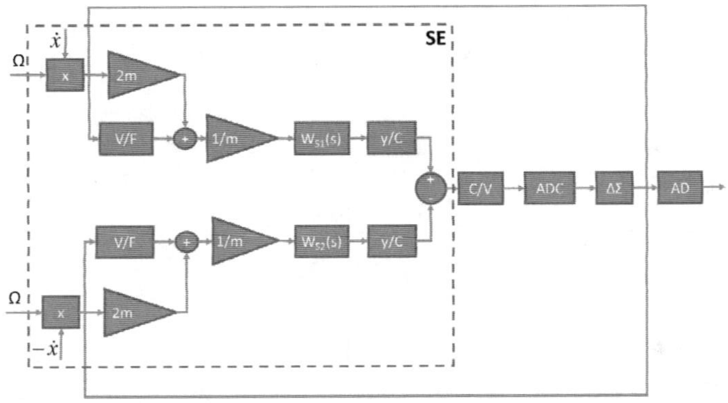

Figure 2. System model of MMG secondary oscillation loop complexing in design process

3.1. Sensing element mechanical part

The transfer functions of SE is formed from the differential equations (1) [16]. Due to technology type determined by possibilities of Elektropribor silicon factory, technical requirements and IC restrictions, the dual-mass architecture of SE with mechanical couple was chosen.

$$\begin{cases} \ddot{x}_{D1} + 2\delta_{D1}\omega_{D1}\dot{x}_{D1} + \omega_{D1}^2 x_{D1} - \omega_{CD}^2 x_{D2} = \dfrac{F_1}{m_1}; \\ \ddot{x}_{D2} + 2\delta_{D2}\omega_{D2}\dot{x}_{D2} + \omega_{D2}^2 x_{D2} - \omega_{CD}^2 x_{D1} = \dfrac{F_2}{m_2}; \\ \ddot{y}_{S1} + 2\delta_{S1}\omega_{S1}\dot{y}_{S1} + \omega_{S1}^2 y_{S1} - \omega_{CS}^2 y_{S2} = -2\Omega_z \dot{x}_{D1}; \\ \ddot{y}_{S2} + 2\delta_{S2}\omega_{S2}\dot{y}_{S2} + \omega_{S2}^2 y_{S2} - \omega_{CS}^2 y_{S1} = -2\Omega_z \dot{x}_{D2}; \end{cases} \tag{1}$$

where x_{Di}, y_{Si} – displacement of *i*-th proof mass along drive (*D*) and sense (*S*) axes (*i*=1..2); δ_{Di}, δ_{Si} – damping ratio of *i*-th proof mass along drive (*D*) and sense (*S*) axes (*i*=1..2); ω_{Di}, ω_{Si} – elastic suspension eigenfrequency of *i*-th proof mass along drive (*D*) and sense (*S*) axes (*i*=1..2); ω_{CD}, ω_{CS} – couple elastic suspension eigenfrequency of *i*-th proof mass along drive (*D*) and sense (*S*) axes; Ω_z – angular rate projection on z axis; m_i - mass of *i*-th proof mass (*i*=1..2); F_i – force, acting on *i*-th proof mass (*i*=1..2).

Equation (1) analysis allows estimating the expressions for resonance in-phase and anti-phase frequencies along drive axis:

$$\omega_{D,in}^2 = \omega_D^2 - \omega_{CD}^2, \tag{2}$$

$$\omega_{D,anti}^2 = \omega_D^2 + \omega_{CD}^2. \tag{3}$$

There are several main ways to organize the couple elastic suspension between masses:

- using springs, directly connecting two or more proof masses [17];
- using springs, connecting two or more proof masse through connection to anchors [8–9, 18–19].

In the second case, stiffness coefficient of couple spring does not influence on anti-phase resonance frequency value but determines the difference between the in-phase and anti-phase resonance frequencies. Some known constructions have such spring only for sense axis [8, 18–19], another one orders to use such connection way for both axes [9].

The confirmation of such effect and the estimation of the expressions for resonance in-phase and anti-phase frequencies with considering stiffness coefficients can be done during the analysis of equation (1) without friction force [20]. The expression for drive axis are the following:

$$\omega_{D1,in}^2 = \frac{1}{m_1}\left(N_1 k_1 - 2\frac{\left(0,5N_1' k_1\right)^2}{\left(N_2 k_2 + N_1' k_1\right)} \right), \tag{4}$$

$$\omega_{D1,\mathrm{anti}}^2 = \frac{1}{m_1}\left(N_1 k_1\right). \tag{5}$$

where k_1 – stiffness coefficient of drive spring, N_1 – number of drive springs; N_1' – number of drive springs, that form couple elastic suspension; k_2 - stiffness coefficient of couple spring along drive axis; N_2 – number of couple springs, that form couple elastic suspension.

The same way the expression for sense axis can be estimated.

IOP Conf. Series: Materials Science and Engineering 1215 (2022) 012010 doi:10.1088/1757-899X/1215/1/012010

3.2. Sensing element transducer

In order to convert the proof mass displacement to capacitance change and to create the electrostatic force that provides the compensation mode, SE includes electrode structures characterized by a displacement-to-capacitance characteristics $C_D(x)$, $C_S(y)$ and voltage-to-force characteristics $F_D(U)$, $F_S(U)$.

At this stage, the model takes into account nonlinear dependencies to describe the «gap-closing» electrode structure of the secondary SE axis loop:

Electrode structure of drive oscillation loop is organized by "comb drive" type (Fig. 3), that has more linear displacement-to capacitance characteristic:

$$C_D(x) = \left(2N_{D,\det} - 1\right)\varepsilon\varepsilon_0 h \frac{L_{D1} + x}{d_{D1}} + 2\left(N_{D,\det} - 1\right)\varepsilon\varepsilon_0 h \frac{L_{D2}}{d_{D2}}, \quad (6)$$

where $N_{D,det}$ – electrode number; ε_0 = 8,854e^{-12} F/m – vacuum permittivity; h – electrode thickness; L_{D1}– electrode overlap length; d_{D1}– gap between fixed and movable electrodes; L_{D2}– gap between electrode end and proof mass or stator; d_{D2}– gap between fixed/movable electrodes.

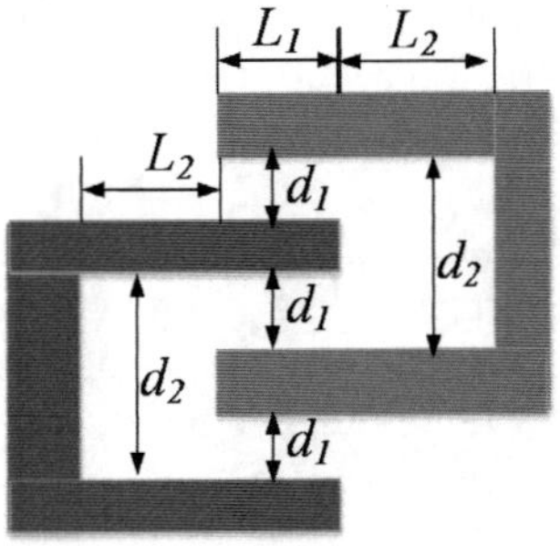

Figure 3. Drive electrode structure

Electrode structure of sense oscillation loop is organized by "gap-closing" type (Fig. 4), that is more sensitive. In order to decrease the nonlinearity of its characteristic a close loop mode is used. Displacement-to capacitance characteristic can be estimated by the following way:

$$C_{S1}(y) = N_{S,\det}\varepsilon\varepsilon_0 h L_S \left(\frac{1}{d_{S1} - y} + \frac{1}{d_{S2} + y}\right), \quad (7)$$

where $N_{S,det}$ – electrode number; L_S– electrode overlap length; d_{S1}– small gap between fixed and movable electrodes; d_{S2}– big gap between fixed and movable electrodes.

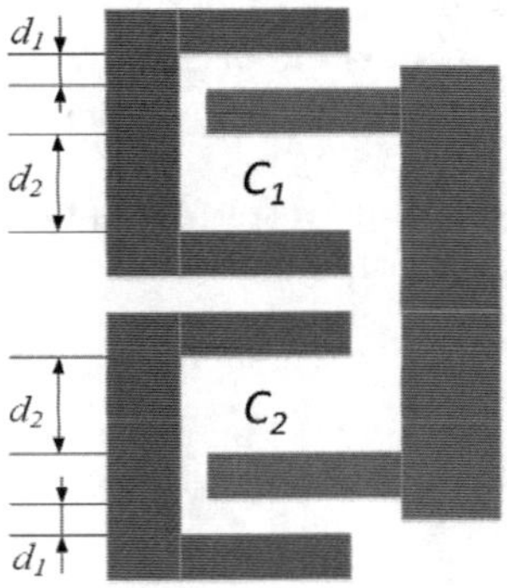

Figure 4. Sense electrode structure

IOP Conf. Series: Materials Science and Engineering 1215 (2022) 012010 doi:10.1088/1757-899X/1215/1/012010

Voltage-to-force characteristic is estimated by the following way:

$$F_i = -\frac{1}{2}\frac{\partial C_{i1}(q)}{\partial q}U^2, \tag{8}$$

where C_{i1} – capacitance of drive or sense electrode structure; q – proof mass displacement along drive or sense axes; U – voltage on electrodes.

3.3. Integrated circuit blocks

MMG operating principle is based on conversion of SE primary vibrational energy into secondary vibrational energy along the axis that is orthogonal to primary oscillation axis. The secondary oscillations contain the information about the measured angular velocity acting along the sensitivity axis that is orthogonal to the primary and secondary axes. Thereby, the integrated circuit of the MMG, working in the compensation mode, has two circuits:

- circuit for measuring capacitance change and motion control along secondary axis;
- circuit for excitation and stabilization of the primary oscillations.

Wherein, the accuracy of primary oscillation amplitude directly affects the accuracy of the sensor.

In order to achieve low noise, wide dynamic range and low non-linearity [7] both circuits are based on digital delta sigma modulator (ΣΔ-M). With time division of sense and drive signals, using ΣΔ-M allows reducing the influence of undesirable nonlinear mechanical effects, nonlinearities in the capacitive electrode structure and electrical parasitic elements due to silicon-on-insulator SE technology.

Wherein, the proof mass of the SE, captured by electrostatic feedback, allows using the mechanical transfer function of the SE as a part of the ΣΔ-M and increasing the order of the delta-sigma modulator by 2 without implementing additional integration units.

Except electronic filters of ΣΔ-M, system model contains the following main blocks:

- capacitance-to-voltage converter, built according to the switched capacitor scheme;
- successive analog-to-digital converter (ADC) with 13-bit resolution
- band pass filter
- amplitude detector
- low pass filter
- proportional-integral-differentiating regulator
- phase adjustment block.

4. Results of system-level modeling

4.1. Output signal spectrums

Figures 5–6 show the spectrums of drive-mode control signal and sense-mode control signal before and after demodulation, obtained by simulation of the refined system model in Matlab/Simulink.

The spectrum forms correspond to the output signals spectrums of the electromechanical delta-sigma modulators with the 11.7 and 11.75 kHz SE eigen frequencies. A signal-to-noise ratio of more than 100 dB is achieved for a bandwidth of 100 Hz. Figures 5 and 6 are obtained at a sampling rate of 400 kHz and a frequency resolution of 1 Hz.

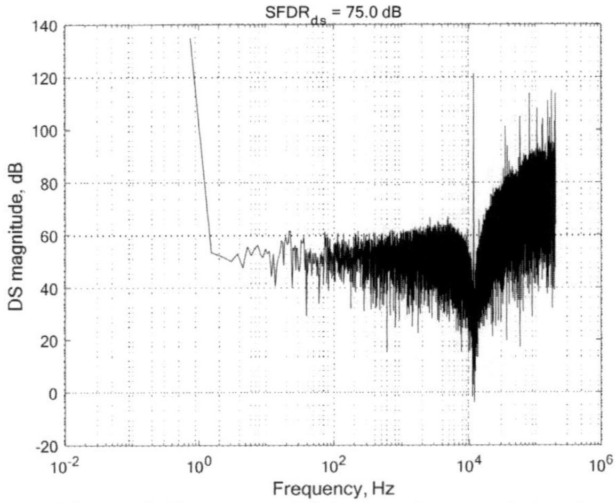

Figure 5. Spectrum of drive-mode control signal

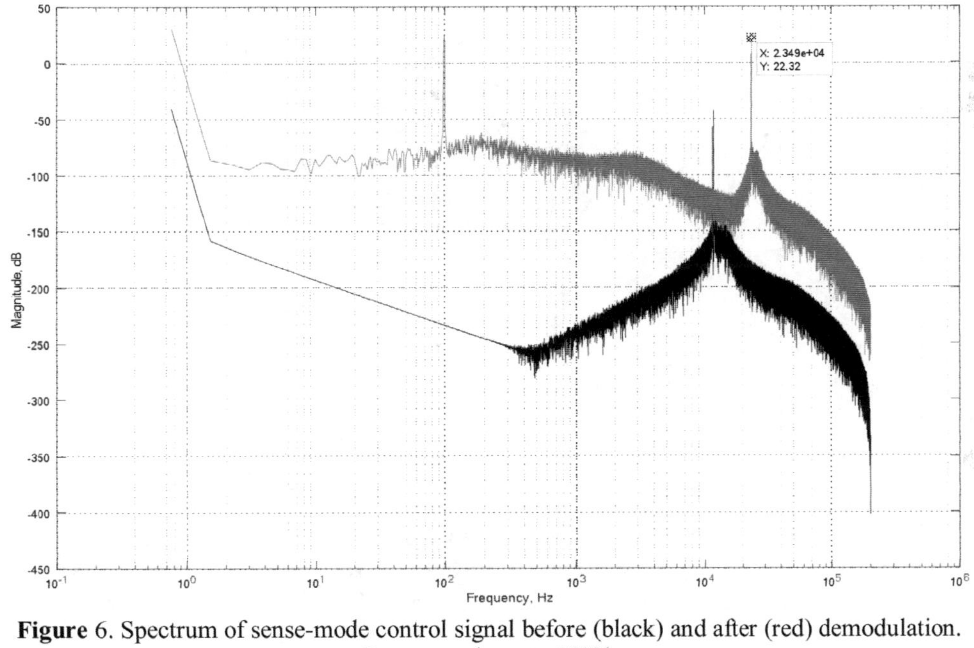

Figure 6. Spectrum of sense-mode control signal before (black) and after (red) demodulation. Input angular rate 450°/s

4.2. SE parameters optimization

The main goals of systems modeling are:

- development of criteria and constraints for further spring optimization;
- estimation of the displacement-to-capacitance characteristics and voltage-to-force characteristics, which describe the electrode structure of SE.

In order to narrow the search area, values of researched parameters are formed considering the following restrictions:

- the gaps d_{D1}, d_{S1}, which determine the sensitivity of the SE, are chosen as minimal as possible, taking into account the structural layer thickness (60 um) and the aspect ratio (20:1);
- the gap d_{S2} is chosen equal to $4d_{S1}$, that is the optimal for two optimization functions: sensitivity maximization and nonlinearity minimization [21];

- the ratio of electrode width and length, in order to avoid damage during manufacture, should not exceed the technological limit 75:1, while the minimum width is 4 um;
- limiting nominal electrodes capacitance, capacity change (for the electrodes of the primary oscillation circuit) and control voltage in accordance with IC restrictions;
- - the minimum sensitivity of the SE along the secondary axis is selected considering the requirements for the sensitivity threshold N_{SE}, reduced to the bandwidth f, and the noise level of the IC N_{IC}:

$$S_{\min} = \frac{N_{ИС}\sqrt{f}}{N_{ЧЭ}}. \tag{9}$$

According to the results of the simulation, response surfaces are constructed (Fig. 7) and the optimal values of the parameters are determined.

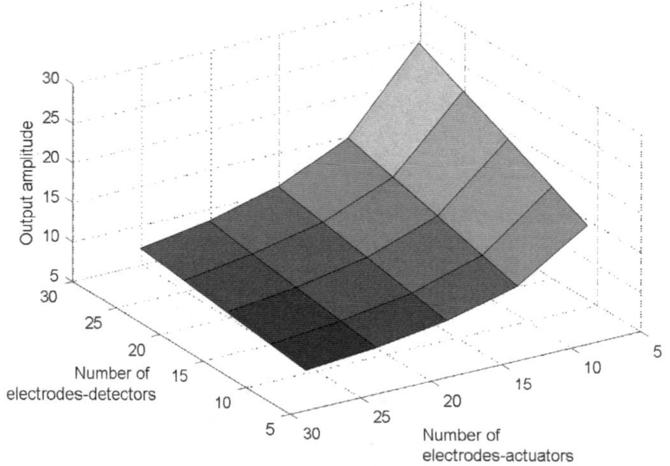

Figure 7. Response surface for output signal amplitude from number of sense-mode electrodes

5. Results of finite element modeling

5.1. Spring optimization

Due to the complex shape of the elastic suspension, which includes a large number of components, the need to ensure the movement of the proof mass along two axes, as well as the anisotropy of the material elastic properties, the first stage of finite element modeling is to optimize the spring elements lengths and widths.

The optimization criteria and constrains estimated at the previous system-level simulation for drive elastic suspension are shown in Table 1.

Table 1. Statement of the problem for drive elastic suspension optimization

	Optimization criteria	Constrains
Drive spring optimization	$f_{D,anti}$ = (11700 ± 25) Hz $\sigma \rightarrow \min$	w = 10 – 15 um L = 300 – 390 um
Couple spring optimization	$k_2{}^* = k_1 \cdot k_2/(k_1 + k_2)$= =286 N/m	w = 7 – 18 um L = 140 – 280 um

* - The formulas are given considering the parallel connection of the couple and drive springs of elastic suspension in the model with one proof mass.

To get accurate results the model for drive spring optimization is a three-dimensional geometry that contains the following construction elements:

- two proof-masses;
- all elastic suspensions;
- all electrodes.

Eigen frequency analysis is used.

Couple spring optimization is carried out in a static statement (determination of displacements under the action of linear acceleration). Thus, the computational model is simplified and represents a quarter of the structure without considering the electrodes.

The calculation was carried out in COMSOL Multiphysics software. According to the results of the calculations, response surfaces are constructed (Fig. 8) and the optimal values of the parameters are determined.

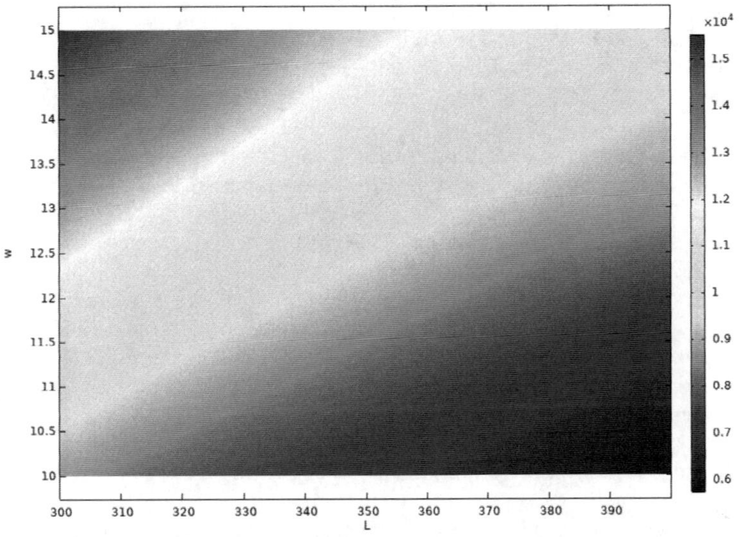

Figure 8. Response surface for anti-phase drive frequency, Hz

5.2. Electrode structure analysis

The analysis of electrical characteristics is carried out with use of a three-dimensional model of three electrodes, two of which are electrically connected. The model must consider the whole technological stack and involve the substrate, silicon oxide layer, structural layer, cap and also cavities. The calculation results are further scaled for the required number of electrodes.

As a result, we have the refined dependences of the capacity on the displacement. The calculation makes it possible not only to refine the characteristics of the electrode structure considering edge effects, but also to optimize the geometry in terms of minimization parasitic parameters.

6. Topology design automotization

Based on the obtained SE geometric parameters, a topology file in the GDSII format was developed for further manufacturing on the silicon factory. To automate the topology generation, the special script was developed. Wherein the script uses the same parameters as specified in the 3D models.

7. Conclusion

The paper presents the complex design method, including simulation at the system level, model refinement based on the results of finite element modelling, and modelling of individual electronic blocks at the circuit level, was described.

The method was used for dual-mass MMG with a measurement range of ± 450°/s design, produced with cavity-SOI technology. The final parameters of MMG SE are presented in Tab.2.

Table 2. Main parameters of dual-mass SE

Parameters	Technical requirement	Drive-mode	Sense-mode
In-phase resonance frequency, Hz	-	9503	10340
Anti-phase resonance frequency, Hz	> 11000	11707	11760
Oscillation amplitude, um	-	0,60	0,05
Displacement to capacity coefficient , pF/um	-	0,384	6,504
Voltage to force coefficient, V^2/N	-	$4{,}79\cdot10^{-8}$	$3{,}33\cdot10^{-7}$

References

[1] Peshekhonov, V.G. The Outlook for Gyroscopy. Gyroscopy Navig., **11**, 193-197 (2020). https://doi.org/10.1134/S2075108720030062.

[2] Yazdi N., Ayazi F., Najafi K. Micromachined Inertial Sensors// Proceedings of IEEE.-1998.- Vol. 86. - №. 8. - P. 1640-1659.

[3] Geen J., Krakauer D. New iMEMS® Angular-Rate-Sensing Gyroscope // Analog Dialog. 2003. Vol. 37. № 3. P. 1–4.

[4] Acar C., Schofield A. R., Trusov A. A., Costlow L. E., Shkel A. M. Environmentally Robust MEMS Vibratory Gyroscopes for Automotive Applications // IEEE Sensors Journal. 2010. Vol. 9. № 12. P. 1895–1906.

[5] Watson J. MEMS Gyroscope Provides Precision Inertial Sensing in Harsh, High Temperature Environments.

[6] Yoon S.W., Lee S., Najafi K. Vibration-induced errors in MEMS tuning fork gyroscope // Sensors and Actuators. 2012. Vol.180. P. 32–44.

[7] Chen F., Li X., Kraft M. Electromechanical Sigma–Delta Modulators Force Feedback Interfaces for Capacitive MEMS Inertial Sensors: A Review //IEEE Sensors Journal. 2016. T. 16. №. 17. P. 6476–6495.

[8] Patent US20160265916A1. MEMS Sensor for Measuring z-Axis Angular Rate.

[9] Patent US 6230563. Dual-mass vibratory rate gyroscope with suppressed translational acceleration response and quadrature-error correction capability.

[10] Lukin A.V., Popov I.A., Belyaev Ya.V., Styazhkina A.V. Nonlinear Dynamics of Micromechanical Wheel Gyroscope under Action of Vibrations//22th Conference of Young Scientists "Navigation and Motion Control – 2020. – pp. 37 - 40.

[11] Eliseev D.P. Povyshenie vibroustoichivosti mikromekhanicheskogo giroskopa RR-tipa. Dis. kand. tekhn. nauk. (Increasing Vibration Resistance of RR-type Micromechanical Gyro. Cand. Eng. Sci. Dissertation. St. Petersburg, 2015.

[12] Moiseev N.V. Mikromekhanicheskii giroskop kompensatsionnogo tipa s rasshirennym diapazonom izmereniya (Torque-to-balance MEMS Gyro with Expanded Range of Measurements). Diss. Cand. Eng. Sc. (2003). St. Petersburg, 2015.

[13] Kovalev A.S. Upravlenie pervichnymi I vtorichnymi kolebaniyami mikromekhanicheskogo giroskopa. (Control of Primary and Secondary Oscillations of MEMS Gyro). Dis. Cand. Eng. Sc. 2008.

[14] Untilov A. A. Issledovanie i razrabotka uprugogo podvesa chuvstvitel'nogo elementa mikromekhanicheskogo giroskopa (Research and Development of Elastic Suspension for MEMS Gyro Sensor): Dis. Cand. Eng. Sc. St. Petrsburg, 2005.

[15] Tulaev A.T., Styazhkina A.V., Kozlov A.S., Belyaev Ya.V. Micromechanical sensors design method based on system-level modeling // Scientific and technical sheets of St. Petersburg State Polytachnic University. Computing, telecommunications and control.

[16] Kempe V. Inertial MEMS. Principles and Practice. CAMBRIDGE University Press, 2011.

[17] Patent US 2010/0313657 A1. Micromachinedtuning fork gyroscopes withultra-high senstivity and shockreection.

[18] Patent US 2010/0139399. Rotation rate sensor.

[19] Patent US 2010/0116050 A1. Coriolis gyro.

[20] Styazhkina A.V., Belyaev Y. V., Belogurov A.A., Bocharov A.N. Sensing element parameter estimation of the dual-mass micromechanical gyroscope using FEM-analysis // 22th Conference of Young Scientists "Navigation and Motion Control – 2020.

[21] Yakimova A.V., Belogurov A.A., Belyaev Y.V. Design of sensing element of micromachined capacitive MEMS accelerometer // International Workshop "Navigation and Motion Control". 2017.

Optimization of a Suspension of a Micromechanical Accelerometer in an Angular Velocity Measurement Unit

I A Iznairov[1,2,3], **M V Efremov**[3], **M A Barulina**[2]

[1]Institute of Precision Mechanics and Control of the Russian Academy of Sciences, Saratov, Russia
[2]Institute of Precision Mechanics and Control of the Russian Academy of Sciences, Saratov, Russia
[3]Research and Production Enterprise Antares, Saratov, Russia

igor@iznairov.ru

Abstract. Optimization of the suspension of the micromechanical accelerometer, which is a part of an angular velocity measurement unit, was considered. This problem was faced because of the need to improve the design of the existing device to meet new technical requirements. Shock loads up to 1 000 g exceed ultimate loads for micromechanical accelerometers by several orders of magnitude. During the study, it turned out that the measurement unit shell works as a resonator, so the shock loads on the micromechanical accelerometer can even exceed 1 000g. In this case, suspensions of the micromechanical accelerometer are deformed in the plane of the accelerometer, which causes their destruction. The solution to this problem was hampered by the strict limits on the overall dimensions of the device. The existing shock dampers that could absorb such shock loads simply did not fit into the attachment points. Therefore, an original spring was designed and manufactured for this purpose. At the final stage, an additional system of passive heat outlets from the accelerometer to the device shell was designed and installed.

1. Introduction

In the work, an angular velocity measurement unit (AVMU) for artificial earth satellites is considered. The main elements of the AVMU are a shell, an interface board, three fiber-optic gyroscopes [1], [2], three micromechanical accelerometers (Fig. 1). AVMU devices with such elements have proven themselves well and therefore they are widely used in different types of satellites [3].

The device under consideration was developed for use in a spacecraft in accordance with some technical requirements. However, after agreeing on the dimensional drawing, the layout of the spacecraft was changed and a new place for AVMU was allocated close to the squib. New allocation led to an increase in the shock loads that the AVMU would experience during its launching into orbit. So, according to the updated technical requirements, the AVMU must withstand up to six shocks with an amplitude of up to 1000g and temperature from -10 to + 50 °C. Updated requirements for shock loading parameters are shown in Table 1.

Table 1. Shock loads parameters.

Frequency range (Hz)	30-50	50-150	150-200	200-500	500-1000	1000-2000	2000-5000
Acceleration (m/s^2)	147-392	392-1570	1570-1962	1962-9810	9810	9810	9810
Acceleration (g)	15-40	40-160	160-200	200-1000	1000	1000	1000
The maximum quantity of shocks - 6							

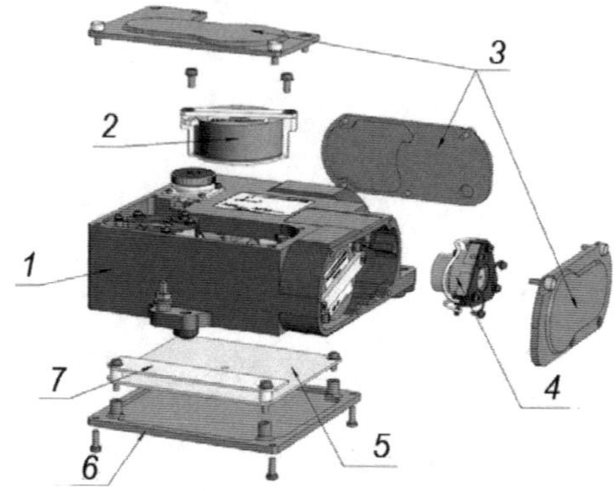

Figure 1. The main elements of the measuring angular velocities unit: unit shell (1), fiber optic gyroscope (2), covers (3), micromechanical accelerometer (4), interface board (5), bottom cover (6), sensor board (7).

2. Full-scale and numerical experiments

To find out how the developed AVMU meets the new requirements, several experiments were carried out on a shaker. These tests showed that after the first impact with an amplitude of 1070g, the device continued to operate, but the output signal was incorrect. After the continuation of the series of shocks, it became obvious that the device was out of order. Numerical simulation (the harmonic response analysis [4]) in Ansys revealed that the AVMU shell worked as a resonator. Therefore, elastic suspensions of a micromechanical accelerometer experience acceleration (1.026·106 m/s2) and stress (8.384·107 Pa) several times higher than the input values of the shock amplitude (9810 m/s2) and allowable stress values [5], [6] (Fig. 2, Fig. 3).

Thus, it was found that the classical fastening of the accelerometer with screws did not absorb shock loads and, accordingly, did not allow us to achieve the required stability of the accelerometer to shock loads.

At the next stage, an analysis of existing shock absorbers was carried out. The shock absorbers had to meet new technical requirements and strict limits on the overall dimensions of the device. We could not find such absorbers. Therefore, an original spring [7] was designed and manufactured for this purpose. (Fig.4a). The use of this spring made it possible to reduce the acceleration and stress on the accelerometer suspensions to acceptable values. (Fig.5).

But the preliminary tests of the AVMU with the original spring for mounting the accelerometer revealed that the physical and geometric properties of the spring did not provide the necessary heat flow from the accelerometer to the device shell. This caused that the accelerometer temperature has been raised to 72 °C. Therefore, at the second stage, an additional system of passive heat removal from the accelerometer to the device body in the form of a heat sink braid was designed and installed (Fig.4b).

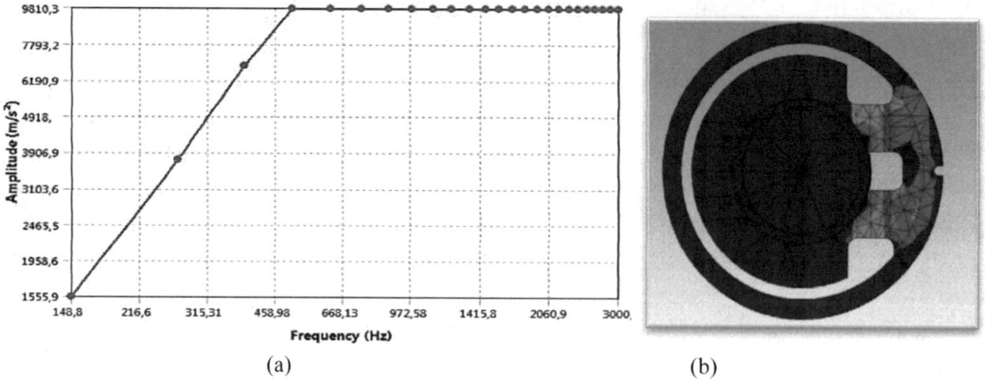

Figure 2. Simulation results: Shock load spectrum (a) and stress-strain state of the accelerometer (b).

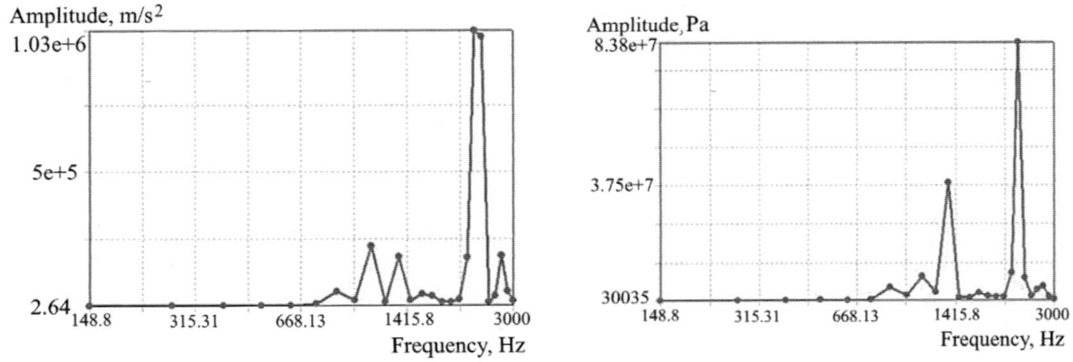

(a) (b)

Figure 3. Harmonic response analysis results for the accelerometer: Acceleration - Frequency response (a) and Stress - Frequency response (b).

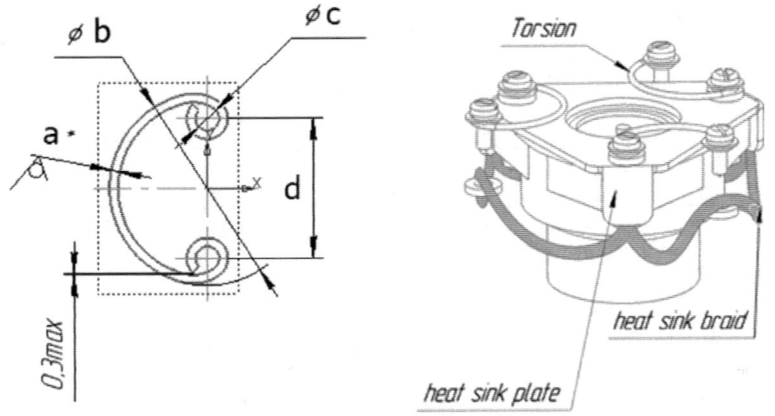

(a) (b)

Figure 4. Cross-section of the designed torsion (a); designed and implemented heat sink braid (b).

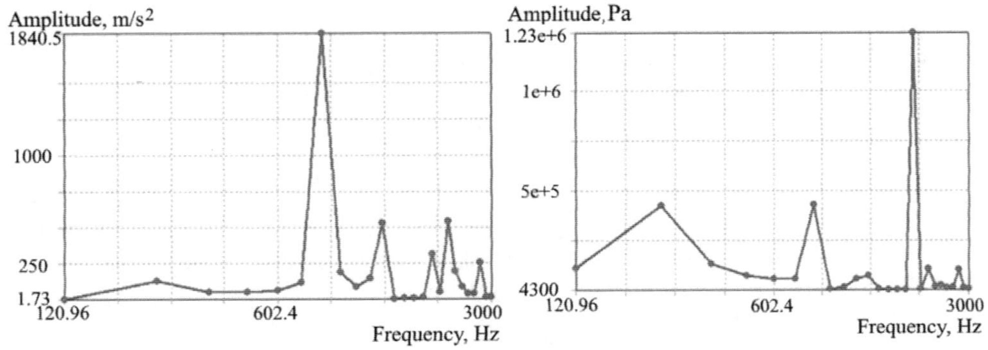

(a) (b)

Figure 5. Harmonic response analysis results for the accelerometer with installed the designed torsion: Acceleration - Frequency response (a) and Stress - Frequency response (b).

Thus, the numerical and full-scale experiments showed that the developed spring and the heat sink braid made it possible to solve the formulated problem. So, with the installed spring and the heat sink braid, shock loads are sufficiently absorbed and the accelerometer temperature does not exceed 62 °C, which is the acceptable value for the accelerometer.

3. Conclusion

As the result of this work, the new suspension of the micromechanical accelerometer, which is a part of the angular velocity measurement unit, was designed and installed to meet new technical requirements.

Moreover, the additional system of passive heat outlets from the accelerometer to the AVMU shell was designed and installed since the physical and geometric properties of the original spring for mounting the accelerometer did not provide the necessary heat flow from the accelerometer to the device shell.

The implemented solution allows the AVMU must remain operational under shock loads up to 1 000 g and thermal effects from – 20 to +50 °C and fully complies with the technical requirements.

References

[1] Lefèvre H. 1993 . The Fiber-Optic Gyroscope. (New York: Artech House)
[2] Korkishko Yu.N., Fedorov V.A., Prilutskiy V.E., Ponomarev V.G., Morev I.V., Obuhovich D.V., Kostritskii S.M., Zuev A.I., Varnakov V.K. and Belashenko A.V. Fiber optic gyro for space applications. Results of R&D and flight tests. 2016 IEEE International Symposium on Inertial Sensors and Systems (USA: Calofirnia pp 37-42.
[3] Golikov A.V., Pankratov V.M. and Efremov M.V. Analysis of Temperature Fields in Angular Velocity Measurement Units on Fiber-Optic Gyros. 2017. Gyroscopy Navig. **9**, 116–123 (2018). https://doi.org/10.1134/S2075108718020050.
[4] Yu Y., Zhang S, Li H/, Wang X and Tang Y. 2017. Procedia Engineering. 174. pp 956–64
[5] Hu S. M. 1982. Journal of Applied Physics 53. p 3576.
[6] Ichiro Yonenaga. 2019. Defects in Crystalline Silicon: Dislocations. Handbook of Photovoltaic Silicon (Springer-Verlag GmbH Germany)
[7] Anuriev V. I. 2001. Handbook of the mechanical engineer 3. pp 180-250

Experimental Study of Nuclear Gyro Bias Compensation Using the Quality Factor Estimation

V.V. Chalkov, A.N. Shevchenko
Concern CSRI Elektropribor, JSC, Saint Petersburg, Russia

rztbro@gmail.com

Abstract. The possibility for bias compensation of nuclear gyro using the quality factor estimation is shown. The corresponding method is described. A description of its application for a nuclear gyroscope in the angle sensor mode is given. The results of experiments confirming the effectiveness of the presented method of nuclear gyro signal refinement are presented.

1. Introduction.

Currently, navigation systems for highly dynamic objects are in demand and, accordingly, interest in the development of small-sized and vibration-resistant rotation sensors and angular velocity sensor has increased. One of the promising directions in the development of a navigation system for highly dynamic objects is a nuclear gyro sensor [1, 2, 3, 4], which is associated with a fundamental insensitivity to vibrations, as well as small weight and size characteristics.

In the course of work with the nuclear gyro developed at the State Research Center of the Russian Federation – Concern CSRI Elektropribor, JSC, a certain zero offset instability was found in the output signal. To solve this problem, we decided to consider the nuclear gyroscope as a self-generation system. Then, parts of the circuit, the schematic representation of which is shown in Figure 1.

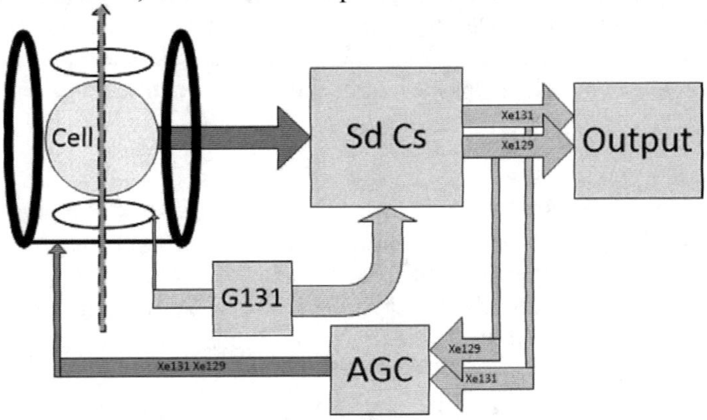

Figure 1. Schematic representation of the self-generation contour of a nuclear gyroscope. Where Sd Cs - block of synchronous detectors and bandpass filters; G131 - block of the generator of the reference frequency Cs; AGC automatic gain control

The following elements are taken [3, 5]: a cell containing xenon isotopes 129 and 131 whose precession frequency and measures a nuclear gyroscope, cesium vapors transferring magnetization to xenon isotopes through spin exchange , a pump laser for optical pumping of cesium vapor; a detection laser for measuring the change in polarization of light as it passes through the cell; a photodetector that converts changes in the polarization of light into an electrical signal; a synchronous detector that separates the total signal of the precession of xenon isotopes; a pair of bandpass filters separating the xenon precession signal 131 and 129 from the total xenon isotope precession signal; magnetic coils that create alternating magnetic fields to maintain the precession of xenon isotopes and automatic gain control (AGC) to maintain the signal level at a given level.

In this consideration, the self-generation frequency of the circuit will be the xenon precession frequency. The zero offset is offset as difference between the self-generation frequency and the Larmor frequency – ω [4, 5].

$$\omega = B \cdot \gamma - \Omega \tag{1}$$

where Ω – angular velocity; B – value of the induction of the magnetic field; γ – gyromagnetic ratio. Considering nuclear gyro as a self-generating system, information on frequency shift can be estimated using information on the change in the quality factor of the circuit. Compensation zero offset instability caused by the shift of the of self-generation frequency will improve the detection accuracy of angular velocity and, accordingly, the determination of the angular velocity.

The aim of this work is to extract the signal from the nuclear gyroscope circuit, which contains information about the self-generation frequency instability, and to develop a bias compensation method using it.

2. Gyroscope zero offset compensation method.

The section of the nuclear gyro self-generation system works as follows. From the precession signals of xenon isotopes using band-pass filters [5,6], information about the change in the frequency of each of the isotopes in the cell is extracted. This signal is applied to coils that create resonant alternating magnetic fields to excite the precession of xenon isotopes in the cell. Thus, a self-generation circuit is formed. The same signal is used to form the value of the output signal about the speed of rotation of the base. For this, a pair of band pass filters and a pair of synchronous detectors are used. The information on the rotation of the base is the frequency shift of precession of Xenon isotopes with different signs. But such a shift can be caused not only by the rotation of the base, but also by a change caused, for example, by the appearance of a magnetic field gradient inside the cell, which will be falsely perceived by the output signal shaping system as a rotation of the base.

As noted earlier the quality factor (Q) of the circuit can also affect frequency shift. Q is a dimensionless parameter that simulates losses in an oscillating structure. It is defined as follows:

$$Q = 2\pi \frac{W_{stored}}{W_{dpp}} = \frac{A_{res}}{A_{stat}} \tag{2}$$

where W_{stored} – amount Energy stored, W_{dpp} – Energy dissipated per cycle, A_{res} – the center-frequency of the peak, A_{stat} – half power or 3 dB bandwidth around the measured peak. And also equation is defined as the ratio of the resonant frequency to the width of the resonant peak.

Currently, there are many methods for monitoring and evaluating the quality factor of an oscillatory circuit [8, 9]. Unfortunately, the existing methods and systems are not suitable for the system under consideration.

The proposed method for evaluating the quality factor for self-generation loop of xenon isotope 131 and quality factor for self-generation loop of xenon isotope 129 based on the operation of AGC, which is already part of the nuclear gyro. AGC's principle of operation is based on maintaining the selected signal level by it's coefficient [10]. Given its presence, equation 2 for both self-generation loop takes the following form:

$$Q_{131} = \frac{A_{stdAGC131}}{C_{AGC131}*A_{stat131}},$$
$$Q_{129} = \frac{A_{stdAGC129}}{C_{AGC129}*A_{stat129}}, \quad (3)$$

where $A_{stdAGC131/129}$ – AGC setpoint for xenon isotope 131/129, $C_{AGC131/129}$ – coefficient AGC self-generation loop of xenon isotope 131/129, $A_{stat131/129}$ – half power or 3 dB bandwidth around the measured peak for xenon isotope 131/129.

In 3 $A_{stdAGC131/129}$ is a given and, accordingly, known value, A_{stat} is also known with sufficient accuracy. From which it follows that information about the quality factor is contained in the AGC coefficient.

Summarizing the above, the following method of bias compensation of nuclear gyro is proposed.. It is necessary to add information about the AGC coefficient to the model of the output signal of the gyroscope using the modified least squares method (MLSM) [11,12].

3. Description of the method.

To test the presented method of zero offset compensation nuclear gyro, an experiment was carried out in which the gyroscope was placed on a fixed base. The self-generating system has been set to the maintenance of a given setpoint level using the AGC. To measure the zero drift of the gyroscope, it worked in the angle sensor mode, this is possible due to the features of the nuclear gyroscope. The measurements were recorded for 66 minutes. The gyroscope drift model can be described as follows:

$$y_i = \alpha_i x_1 + \beta_i x_2 + t_i^2 x_3 + t_i x_4 + T_i x_5 \quad (4)$$

where i - measurement number, y_i - gyroscope measurements at time i, x_1, x_2, x_3, x_4, x_5 - unknown coefficients to be determined during calibration, α_i – xenon isotope precession frequency 131, β_i – xenon isotope precession frequency 129, t_i – time, T_i – pump laser temperature.

To expand the gyroscope drift model, it is proposed to use information about AGC coefficient, in the following form:

$$I_{i131} = \sum_{j=0}^{i} C_{AGC131,j} \quad (5)$$

$$I_{i129} = \sum_{j=0}^{i} C_{AGC129,j} \quad (6)$$

Then, taking into account 5.6, equation 4 will take the following form:

$$y_i = \alpha_i x_1 + \beta_i x_2 + t_i^2 x_3 + t_i x_4 + T_i x_5 + I_{i131} x_6 + I_{i129} x_7 \quad (7)$$

Proceeding from the fact that the gyroscope is installed on a fixed base, then the measured angle is set equal to 0 (the speed of the Earth's rotation can be neglected). Taking into account (7), the measurement equation can be written in the following form:

$$Y_{m\times1} = H_{m\times7} x_{7\times1} + v_{m\times1} \quad (8)$$

where v – Gaussian white noise measurements, x – a state vector of the form $x = [x_1\ x_2\ x_3\ x_4\ x_5\ x_6\ x_7]^T$, $Y = [y_1\ y_2 \dots y_m]^T$, H – measurement matrix and has the following form:

IOP Conf. Series: Materials Science and Engineering 1215 (2022) 012012 doi:10.1088/1757-899X/1215/1/012012

$$H_{m\times 7} = \begin{bmatrix} \alpha_1 & \cdots & \alpha_m \\ \beta_1 & \cdots & \beta_m \\ t_1^2 & \cdots & t_m^2 \\ t_1 & \cdots & t_m \\ T_1 & \cdots & T_m \\ I_{1,131} & \cdots & I_{m,131} \\ I_{1,131} & \cdots & I_{m,131} \end{bmatrix}^T, \; m - \text{measurement number.} \quad (9)$$

Using MLSM [9], one can obtain estimates for the vector x. Then, using the estimate of the vector x, we obtain an approximation of the measurements $\hat{Y} = H\hat{x}$ the residual of measurements is built:

$$g = Y - H\hat{x} \quad (10)$$

where $\hat{x}$ – vector x estimate.

4. Experimental data processing results.

To demonstrate the presented method, plots of g were plotted as for the gyroscope drift model without information about AGC coefficient 4 Figure 2 (left), and using AGC coefficient information 7 Figure 2 (right). To evaluate the obtained experimental data, graphs of the Alan variation versus g were plotted, shown in Figure 3.

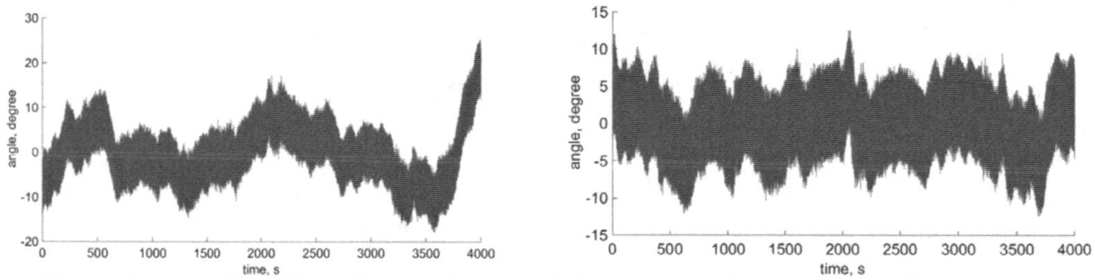

Figure 2. zero offset graph of the nuclear gyro left) Without using AGC coefficient information in model right) Using AGC coefficient information in the signal model.

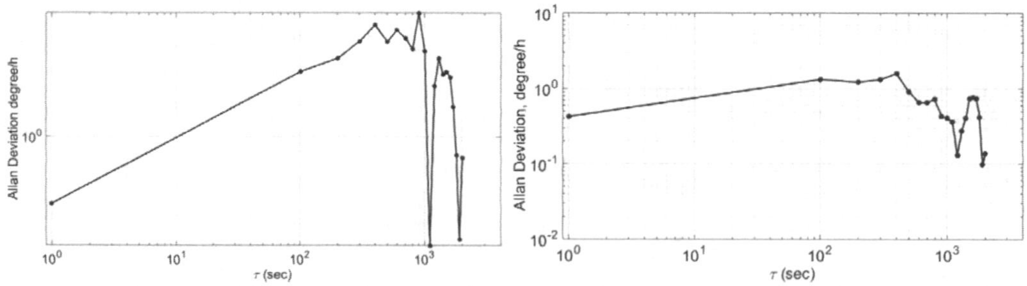

Figure 3. graph of Alan deviation zero offset graph of the nuclear gyro left) Without using AGC coefficient information in model right) Using AGC coefficient information in the signal model.

Analyzing the obtained graphs, it can be seen that in Figure 2 the right-hand graph of the zero offset of the nuclear has a smaller amplitude than the left one. This is confirmed by the Alan variation plots in Figure 3. From which we can come to the conclusion that the use of the nuclear gyro output signal in the model as a correcting Q-factor, based on the AGC coefficient value, leads to an increase in the nuclear gyro accuracy.

5. Conclusions.

The paper shows the possibility of using the quality factor for nuclear gyro bias compensation. The corresponding method is described. A description of its application for a nuclear gyroscope in the angle sensor mode is given. The results of experiments confirming the effectiveness of the presented method of nuclear gyro signal refinement are presented. In the future, a more detailed study of the reasons for the instability of the figure of merit is expected.

References

[1] Peshekhonov, V.G. The Outlook for Gyroscopy. Gyroscopy Navig. 11, 193–197 (2020).

[2] Loesche, A., Nuclear Induction [Russian translation], Moscow: Inostrannaya literatura, 1963.

[3] Vershovskii, A.K., Litmanovich, Yu.A., Pazgalev, A.S. and Peshekhonov, V.G., Nuclear magnetic resonance gyro: ultimate parameters, Gyroscopy and Navigation, 2018, vol. 9, no. 3, pp. 162–176.

[4] Aleksandrov, E.B. and Vershovskii, A.K., Modern radio-optical methods in quantum magnetometry, Physics–Uspekhi, 2009, vol. 52, no. 6, pp. 573–601.

[5] Ding, M., Development and prospects of quantum sensing technology, Proc. 20th Conference of Young Scientists "Navigation and Motion Control", Concern CSRI Elektropribor, St. Petersburg, 2018, pp. 21–22.

[6] Riehle, F., Frequency Standards: Basic and Applications [Russian translation], Moscow: Fizmatlit, 2009.

[7] Farrar, T.C. and Becker, E.D., Pulse and Fourier Transform NMR: Introduction to Theory and Methods, New York: Academic Press, 1971.

[8] J. Han et al., "Electrostatic stabilization of thermal variation in quality factor using anchor loss modulation," SENSORS, 2014 IEEE, 2014, pp. 998-1001, doi: 10.1109/ICSENS.2014.6985171.

[9] B. Kim et al., "Temperature Dependence of Quality Factor in MEMS Resonators," in Journal of Microelectromechanical Systems, vol. 17, no. 3, pp. 755-766, June 2008, doi: 10.1109/JMEMS.2008.924253.

[10] Bezmen, G V; Shevchenko, A N. Journal of Physics: Conference Series; Bristol Volume 1864, Ed. 1, (May 2021). DOI:10.1088/1742-6596/1864/1/012010, «The Choice of Criteria and Methods to Estimate the Quality of the Angular Sensor Based on the Effect of Nuclear Magnetic Resonance».

[11] Baryshev, A.V. and Fedotova, E.F., Finding optimal solutions for linear programming decision problems using Microsoft Excel Solver Add-in, Naukovedenie Internet Journal, 2015, vol. 7, no. 3(28), p. 88.

[12] Stepanov O.A. Estimation algorithms for navigation data processing [Russian translation], Metodichescoe posobie. - St. Petersburg: ITMO University, 2021. - 174 p.

IOP Conf. Series: Materials Science and Engineering 1215 (2022) 012013 doi:10.1088/1757-899X/1215/1/012013

Development of Alkali Vapor Cell for Miniature Nuclear Magnetic Gyroscope

M A Vlasova[1,2], A N Shevchenko[1]

[1] Concern CSRI Elektropribor, JSC, Saint Petersburg, Russia
[2] ITMO University, Saint Petersburg, Russia

mary_vlasova446@mail.ru
standw.shev@gmail.com

Abstract. The article presents numerical requirements for vapor cells for a nuclear magnetic gyroscope. The design requirements for the cells, such as shape, size and transparency, as well as their influence on the characteristics of the gyroscope, in particular, on the angle random walk, are considered. The process of manufacturing vapor cells and control of their parameters is described.

1. Introduction

The nuclear magnetic gyroscope (NMG) is a quantum sensor whose principle of operation is based on the interaction of light and matter [1–6]. The working substances of the gyroscope, as well as the buffer gas, are located in a cell with alkali metal vapors [7–9].

There are a number of requirements for such a cell, in particular:

- the cell material should not react with alkali metal atoms;
- the cell should pass the laser radiation of pumping and detection;
- the cell should be sealed, heat-resistant and non-magnetic;
- the shape and dimensions of the cell should serve the technical requirements after its sealing.

The cells are filled with a mixture of gaseous substances. The composition of this mixture includes the following components: alkali metal vapors, noble gas, buffer gas. An anti-relaxation coating can be applied to the inner walls of the cell. Cesium or rubidium is used as an alkali metal. Nuclear magnetic resonance is excited in a noble gas, usually in xenon isotopes ^{129}Xe and ^{131}Xe. Nitrogen N_2 and a natural mixture of xenon are used to absorb the secondary radiation of the gas mixture, as well as a buffer gas to prevent collisions of alkali metal vapors with the cell walls. The process of filling vapor cells is a complex technical task.

When developing the requirements for the design of the vapor cell, the influence of its parameters on the characteristics of the gyroscope, in particular, on the angle random walk (ARW), should be analyzed.

The purpose of this work is to form requirements for a sapphire vapor cell for a miniature nuclear magnetic gyroscope and to develop its design and technologies for its manufacture, filling and parameter control. Experimental approbation of technologies for filling and control of parameters of vapor cells was conducted on a prototype of a cell made of glass.

2. Dependence of ARW on cell parameters

Main components of a gyro error are:

- quantization noise,
- bias instability (or bias drift),
- angle random walk.

Quantization noise is mainly due to analog-to-digital conversion of the gyro output signal. Bias instability, expressed in °/s or °/h, is the peak-to-peak amplitude of the bias long term drift. Angle random walk (ARW) is a noise contribution to the rotation angle value obtainable by integrating the angular velocity. W_{ARW} is the angle random walk coefficient, usually expressed in °/√s or °/√h, which shows standard deviation of noise observed in rotation angle estimation [10–12]. It can be written as:

$$W_{ARW} = k_{ARW} \cdot \sigma(\omega)\big|_{1Hz},$$ where:

$k_{ARW} = \dfrac{3600 \cdot 360}{2\pi} \cdot \dfrac{1}{\sqrt{2} \cdot 60}$, $\sigma(\omega)\big|_{1Hz} = \delta W_{ARW} = \dfrac{1}{Q} \cdot \sqrt{\Delta f}$ – standard deviation of noise observed in rotation angle estimation with bandwidth, Q - quality factor.

The Q factor is determined by the signal-to-noise ratio and the relaxation rate of the working substance [13–14]:

$$Q \approx \frac{S}{\sigma_N \Gamma},$$

where: S – signal amplitude; σ_N – spectral noise density; Γ – spin relaxation rate.

Thus, it is possible to estimate the ARW of the gyroscope by obtaining the ratio of the signal amplitude to the noise spectral density and the spin relaxation rate. Next, let's consider the dependence of these three factors determining the ARW of the NMG on the size of the vapor cell.

In NMG, the signal amplitude is determined by the amplitude in the xenon precession signal processing channel. The spectral density of noise, in turn, is determined by noise in the xenon channel and noise in the alkali metal channel.

The cell size affects the relaxation rate of the alkali metal Γ (cesium or rubidium), as well as the rate of xenon relaxation on the cell walls. Xenon relaxation occurs during collisions with alkali metal atoms, but at the same time, under the action of optical pumping, polarization of alkali metal atoms occurs, which polarizes xenon atoms through spin exchange. The achievable degree of polarization of xenon atoms depends on the degree of polarization of the alkali metal. The degree of polarization of xenon determines the signal and noise levels in its channels, and the polarization and relaxation rate of the alkali metal, in turn, determine the intrinsic noise in the channel of the alkali metal. Thus, the signal and noise in the xenon channels together with the noise in the alkali metal channel affect the ARW of the NMG. The algorithm for calculating the dependence of ARW on the linear cell size described above is graphically shown in Fig. 1. The parameters determining the intensity of optical pumping and the loss of polarization during collisions of xenon isotopes with alkali metal atoms do not depend on the cell size and were set by constants.

As a result of calculations using the above algorithm for calculating the dependence of ARW on the linear cell size, the dependence of the limiting ARW of the NMG on the size of its vapor cell is obtained. It is presented in Fig. 2.

The graph is obtained taking into account the capacities of commercially available vertical radiation lasers (0.3 mW). The graph shows that the minimum ARW is achieved with a cell size from 3 to 8 mm.

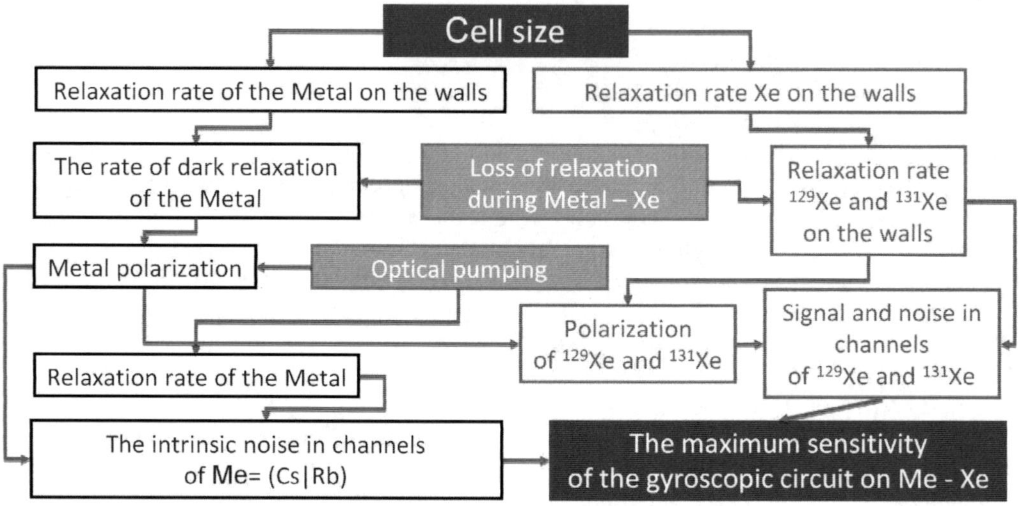

Figure 1. Algorithm for calculating the dependence of the ARW on the linear size of the cell.

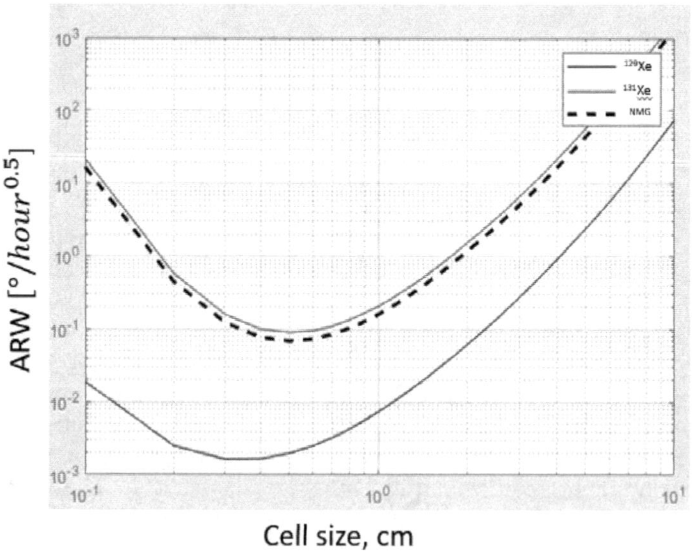

Figure 2. Graph of the dependence of the gyroscope ARW on the cell size.

3. Design requirements for cells

We formulate a number of additional requirements for the design of the vapor cell, in addition to the dimensions.

The material for vapor cells should be transparent enough to effectively transmit laser light, and should not interact with alkali metal vapors. Borosilicate glass (B_2O_3) is usually used to create gyroscope cells. Borosilicate glass is a silicate glass in which the alkaline components in the feedstock are replaced by boron oxide (differs from ordinary glass with increased heat resistance and increased resistance to mechanical damage).

The shape of the cell should be symmetrical – spherical or cubic.

A cell size of 5 mm was selected for the production of prototypes.

For miniature cells with a characteristic size (diameter or edge length) of less than 10 mm, a cubic cell shape is preferred, since a spherical miniature cell introduces unacceptable optical distortions. Spherical cells with a diameter of more than 15 mm do not introduce noticeable optical distortions [15, 16].

The most popular technology for creating cubic small-sized cells is the technology of melting and laser soldering of glass [17, 18], and spherical cells are made using glass blowing technology [19, 20], which, however, does not allow the production of small-sized cells. The disadvantage of miniature cubic cells is the difficulties associated with the creation of flat faces and the preservation of a given geometry, as well as with the rounding of the faces when melting glass.

The disadvantage of cells blown out of glass is the occurrence of optical distortions in the light transmitted through them and the difficulty of ensuring the repeatability of the cell shape. A method of manufacturing cells from sapphire and glass is known [21]. Such cells combine excellent optical properties and ease of sealing due to the desoldering of the glass tube.

Unlike the cells described in the article [21], this paper proposes a design consisting of a sapphire cell body and a glass tube for filling and sealing. The cell body is ground out of a solid sapphire blank and has no component parts.

In the process of manufacturing cells, it is required to connect only 2 parts of the structure: the body of the sapphire cell and the glass tube, which is attached to its lower base. The design of the cubic cell of a nuclear magnetic gyroscope is shown in Fig. 3.

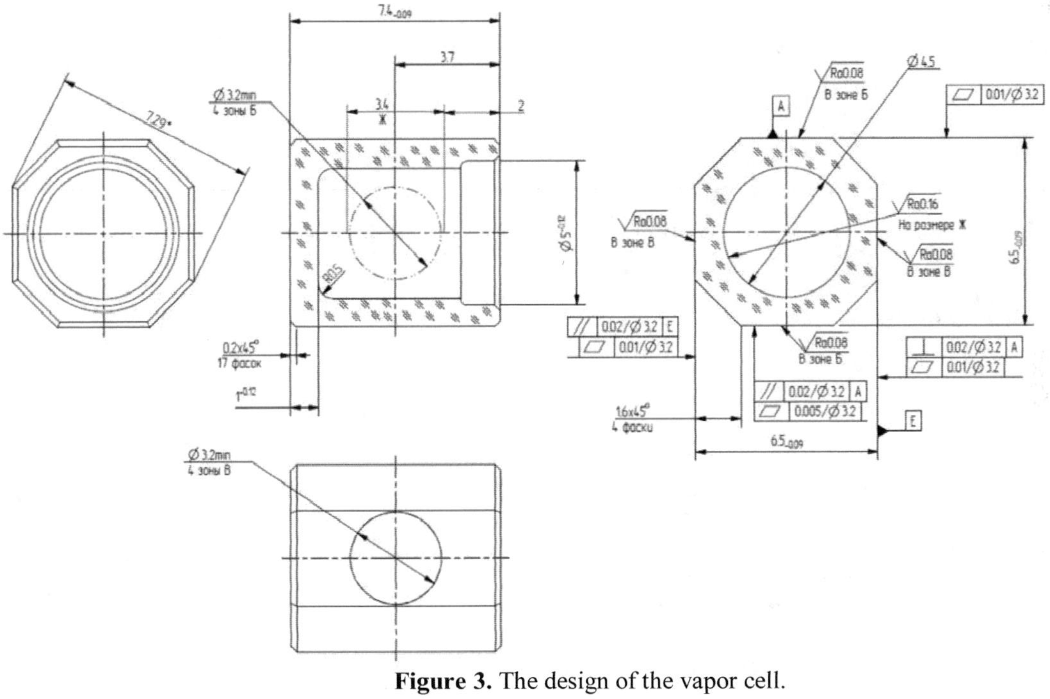

Figure 3. The design of the vapor cell.

Currently, the technology of connecting the structural parts of a cubic cell is being studied. When trying to get a glass-sapphire connection, difficulties arose. The resulting cell prototypes turned out to be leaky, which was evident by the absence of optical resonance signals in them and by the color change of the alkali metal due to oxidation. To visualize the defects of the sapphire and glass adhesions, an experiment was performed, which consisted in the following. A poly-para-xylylene coating was applied to the walls of the test cell using a vacuum installation designed for applying protective coatings to electronic printed circuit boards [22]. It is optically transparent, so the cells were annealed at 400 °C. After annealing, the coating became contrasting – it became possible to assess the quality of the connecting seams. After annealing, the existing soldering defects (Fig. 4) of the transparent elements of their construction became visible, which indicates an insufficient level of tightness. Work on obtaining a high-quality glass-sapphire compound will continue.

Technologies for filling cells, as well as monitoring their parameters, were worked out on spherical cells obtained by blowing glass.

Figure 4. Experiment on visualization of soldering defects of transparent cell parts

4. Filling the cells with a gas mixture

We describe the process of filling the vapor cells with working substance.

The installation used to fill the NMG prototype vapor cells included four important elements: a vacuum station, a metal-glass junction, a gas intake system and a pressure control system (Fig. 5).

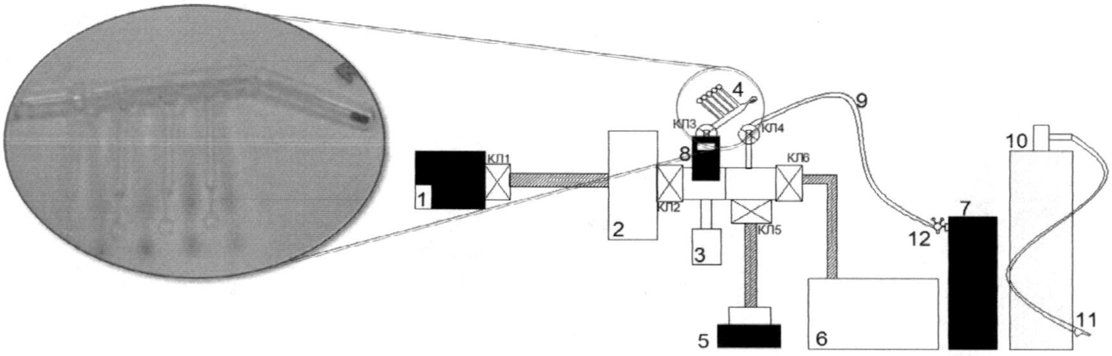

Figure 5. Connection diagram and mutual arrangement of the vacuum installation for filling the cells with gas. 1– pre–vacuum pump; 2 – turbomolecular pump; 3 – low pressure sensor with hot Pirani/Bayard–Alpert cathode; 4 – tripod with NMG; 5 – vacuum ion pump; 6 – helium leak detector; 7 – gas cylinder with nitrogen; 8 – piezoelectric high pressure sensor; 9 – copper tube for nitrogen supply to the system; 10 – gas cylinder with helium; 11 – gas gun for blowing the installation with helium; 12 – gas reducer.

Nitrogen gas with a volume fraction of pure nitrogen of at least 99.999% was used to fill a glass vapor cell. A glass holder with hollow vapor cells was used as a blank, inside each of which there is a tablet with cesium salts. The vacuum installation for filling the cells with gas was prepared according to the instructions provided: the vacuum fittings were washed and dried, the installation was assembled according to the scheme, the entire system was checked for tightness using a helium leak detector, vacuum was created, and the pressure inside the installation was monitored. A cell with steam was obtained after performing the following actions: a tube was washed from the corresponding glass of the desired diameter; a cell of the required diameter was obtained from the tube by blowing glass, with a given wall thickness and tube length. The cell is connected to a vacuum installation; air is pumped out of the cell to a pressure of 10-3 torr; the cell is filled with nitrogen to a pressure of about several torr; the discharge ignites; nitrogen is pumped out after 20 minutes. The cell was continuously heated to a temperature of 500-550°C with continuous pumping to 10-5 torr. Cleaning of the inner surface of the cell and tube by discharge in nitrogen and xenon at a gas pressure of the order of several torr was repeated. Next, the cuvette was filled with metallic cesium, for this: an ampoule with cesium salts was heated to a temperature of 600-700°C, preventing the glass from softening, the reduced cesium was distributed through the cuvettes using a gas burner for cesium distillation. At the final stage, the cell was cooled, filled with the necessary mixture of nitrogen and xenon and soldered out of the installation.

5. Cell control based on optical absorption experiment

The control of the vapor cells was carried out using a special installation for monitoring the cells (Fig. 6). It was assembled as follows: the laser emitter and the photodetector are mounted on the same axis, the vapor cell installation site is mounted according to a thermostabilized scheme so that the laser beam can pass through the cell, an oscilloscope is used for receiving the photodetector information signal.

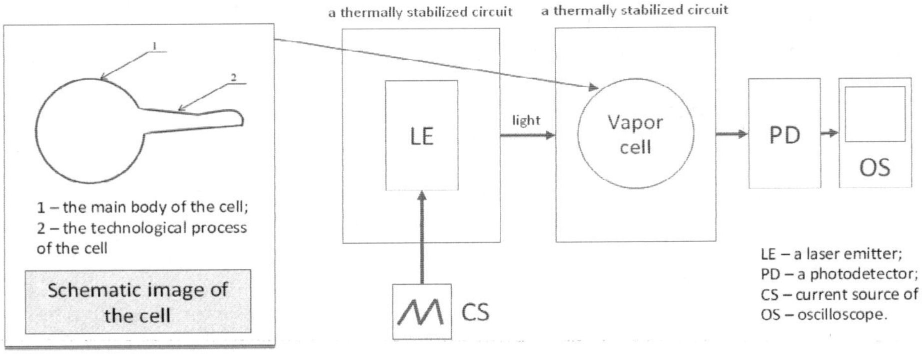

Figure 6. Connection diagram and mutual arrangement of the elements of the equipment for testing the cell with the help of an optical absorption experiment.

A vertical-cavity surface-emitting laser [23, 24] was used as a laser emitter, which was connected to a current source (Rigol DG1022 function generator). The laser's DC bias current was changing from 2 to 3 mA to sweep the laser frequency near the D1 133Cs line. Figure 7 shows the optical absorption spectrum. In a good sealed cell, the two absorption lines are clearly observed at temperature above 60°C and can be associated with the ground state hyperfine splitting in cesium (F=3, F=4). Note that inhomogeneous line broadening due to collisions between cesium and buffer and impurity gas atoms limits the resolution of the D1 133Cs transitions. If there are no absorption lines, it means that the cell is not hermetic and oxygen from the ambient air has penetrated into the cell and oxidized cesium.

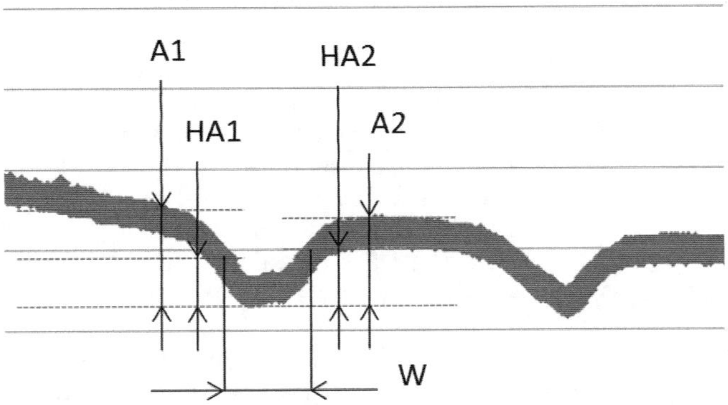

Figure 7. Width of the line of alkali metal optical absorption.

6. Conclusion

The requirements for the cell design are formulated and its dimensions are determined, providing a minimum of the gyroscope ARW. The design of the sapphire body of the vapor cell has been developed and a method for visualizing defects in the connection of transparent structural elements has been proposed and experimentally tested. The filling technology has been tested on a prototype of a glass cell. The quality of the manufactured cell was experimentally estimated using an optical absorption experiment.

In the future, it is planned to master the methods of connecting sapphire and glass and to hold experimental studies of sapphire cells.

References

[1] Peshekhonov, V.G. The Outlook for Gyroscopy. Gyroscopy Navig. 11, 193–197 (2020). https://doi.org/10.1134/S2075108720030062

[2] Walker T G, Larsen M S 2016 Spin-exchange-pumped NMR gyros. *Advances in atomic, molecular, and optical physics* **Vol. 65.** p 377-405 https://doi.org/10.1016/bs.aamop.2016.04.002

[3] Litmanovich YU A, Vershovskij A K, Peshekhonov V G 2014 Giroskop na osnove yavleniya yadernogo magnitnogo rezonansa: proshloe, nastoyashchee, budushchee Materialy plenarnogo zasedaniya 7-j Rossijskoj mul'tikonferencii po problemam upravleniya, AO «Koncern «CNII «Elektropribor» p 35-42 (in Russian)

[4] Umarkhodzhaev R M, Pavlov Y V, & Vasil'ev A N 2018 History of NMR Gyroscope Development in Russia in 1960–2000s *Gyroscopy and Navigation* **9(3)** p 147-161 DOI: 10.1134/S2075108718030094

[5] Vershovskij A K, Shevchenko A N 2015 Yadernyj magnitnyj giroskop: princip dejstviya, istoriya, perspektivy Navigaciya i upravlenie dvizheniem p 19-28 (in Russian)

[6] Bezmen G V, Shevchenko A N, Kostin P N, Berzeitis A N, Bezmen V S, & Petrov V I 2020 A two-frequency spin oscillation system for a quantum angular rate sensor *Gyroscopy and Navigation* **11(2)** p 115-123 DOI: 10.1134/S2075108720020030

[7] Shevchenko, A. N., & Zakharova, E A 2019 Research of magnetic resonance quality factor dependence on cesium distribution in quantum rotation sensor cell *Journal Scientific and Technical Of Information Technologies, Mechanics and Optics* ***122*(4)** p 567 DOI: 10.17586/2226-1494-2019-19-4-567-573

[8] Shevchenko A N, Kuzmin, A G, & Titov Yu A, 2018 Mass-spectrometric measurement of the gas mixtures composition in the cells of a quantum rotation sensor. *Nauchnoe priborostroenie*, **28(2)**, p 62-68.

[9] Popov E N, Barantsev K A, Ushakov N A, Litvinov A N, Liokumovich L B, Shevchenko A N, ... & Medvedev A V 2018 Behavior of signal from optical circuit of quantum rotation sensor based on nuclear magnetic resonance *Gyroscopy and Navigation* **9(3)** p 183-190 DOI: 10.1134/S2075108718030082

[10] Vershovskii A K, Litmanovich Yu A, Pazgalev A S, Peshekhonov V G 2018 Nuclear magnetic resonance gyroscope: the ultimate parameters *Gyroscopy Navig.* **9(3)** p 162-176 DOI: 10.1134/S2075108718030100

[11] IEEE Std 1554-2005. IEEE Recommended Practice for Inertial Sensor Test Equipment, Instrumentation, Data Acquisition, and Analysis. Approved 5 October 2005, Reaffirmed 16 April 2012 by American National Standards Institute. IEEE Aerospace and Electronic Systems Society. Chapter 19.7.2 – Velocity and angle random walk per root hour.

[12] Armenise M N, Ciminelli C, Dell'Olio F, & Passaro V M 2010 *Advances in Gyroscope Technologies* (Berlin: Springer-Verlag)

[13] Pomerancev N M, Ryzhkov V M, Skrockij G V 1972 *Fizicheskie osnovy kvantovoj magnitometrii* (Moskva: Nauka) (in Russian)

[14] Vershovskii, A K & Pazgalev A S 2008 Optimization of the Q factor of the magnetic M x resonance under optical pump conditions *Technical Physics* ***53*(5)** p 646-654 DOI: 10.1134/S1063784208050198

[15] Gubin V P, Otrohov S YU 1989 Vliyanie gazovoj linzy na formu uzkogo rezonansa v yachejke i interferometre Kvantovaya elektronika. T. 16. №. 12. P 2573-2580 (in Russian)

[16] Akul'shin A M, Velichanskij V L, ZHerdev A I, Zibrov A S, Malahova V I, Nikitin V V, ... & Harisov G G 1989 Selektivnoe otrazhenie ot granicy steklo–gaz pri bol'shih uglah padeniya sveta Kvantovaya elektronika T. 16. №. 3. p 631-637 (in Russian)

[17] Pat. 2 554 358, Rossijskaya Federaciya, MPK C03B 23/20, G04F 5/14. Sposob izgotovleniya malogabaritnyh atomnyh yacheek s parami atomov shchelochnyh metallov i ustrojstvo dlya ego osushchestvleniya / Fishman R. I., Sevost'yanov D.I., Vasil'ev V. V., Zibrov S. A., Sivak A. V., Velichanskij V. L., № 2018115563, zayavl.

2018.04.25, opubl. 2019.03.28 (in Russian)

[18] Pat. 2 677 154, Rossijskaya Federaciya, MPK C03B 23/18, B23K 26/21. Sposob izgotovleniya malogabaritnyh atomnyh yacheek s parami atomov shchelochnyh metallov / Egorov A.B., CHuchelov D. S., Fishman R. I., Velichanskij V. L., № 2018115561, zayavl. 2018.04.25, opubl. 2019.01.15 (in Russian)

[19] Knappe S, Velichansky V, Robinson H G, Kitching J, Hollberg L 2003 Compact atomic vapor cells fabricated by laser induced heating of hollow-core glass fiber *Rev. Sci. Instrum.* ***74*(6)** 3142-3145 https://doi.org/10.1063/1.1575925

[20] Sevost'yanov, D I, Sivak A V, Losev S S, Mas'yan YU V, Fishman R I, Vasil'ev V V, ... & Velichanskij V L 2015 Metod izgotovleniya malogabaritnyh atomnyh yacheek i ih metrologicheskie harakteristiki. Elektromagnitnye volny i elektronnye sistemy 20(8) p 73-81

[21] Sarkisyan D G 1988 Kvantovaya elektronika T. 15 №11 2358 (in Russian)

[22] Vlasova M A, SHevchenko A N 2021 Gazovaya yachejka s antirelaksacionnym pokrytiem dlya kompaktnogo magnitometra dlya zadach biomagnitnyh issledovanij Sbornik tezisov dokladov kongressa molodyh uchenyh. Elektronnoe izdanie. Elektronnyj resurs (in Russian)

[23] Bobrov M A, Blokhin S A, Maleev N A, Blokhin A A, Vasyl'Ev A P, Kuzmenkov A G, ... & Ustinov V M 2019 Effect of coherent population trapping in a compact microfabricated Cs gas cell pumped by intra-cavity contacted VCSELs with rhomboidal oxide current aperture. *In Journal of Physics: Conference Series* ***1400 (7)*** p 077014 IOP Publishing DOI: 10.1088/1742-6596/1400/7/077014

[24] Maleev N A, Blokhin S A, Bobrov M A, Kuz'menkov A G, Kulagina M M, & Ustinov V M 2018 Laser Source for a Compact Nuclear Magnetic Resonance Gyroscope *Gyroscopy and Navigation* ***9(3)*** p 177-182 DOI: 10.1134/S2075108718030057

IOP Conf. Series: Materials Science and Engineering 1215 (2022) 012014 doi:10.1088/1757-899X/1215/1/012014

Estimation of the Nuclear Magnetic Gyroscope Sensitivity to Inertial Influences

A.S. Zavitaev, M. I. Evstifeev

ITMO University, Saint Petersburg, Russia
Concern CSRI Elektropribor, JSC, Saint Petersburg, Russia

zav.al98@mail.ru

Abstract. The factors influencing the stability of the zero offset of a nuclear magnetic gyroscope (NMG) are considered. As a result, an equation for the connection of these parameters with changes arising in the course of inertial influences on the NMG body has been created. In addition, the displacements caused by vibration effects on the NMG under operating conditions are analyzed. Based on these data, the changes in NMG accuracy are calculated and conclusions are drawn about the effectiveness of the NMG design.

1. Introduction

The paper deals with the nuclear magnetic gyroscope (NMG) which being developed at the Concern CSRI Elektropribor. This gyroscope is supposed to be installed on highly dynamic objects. During its operation, NMG is inevitably subjected to various physical disturbances, which results in deterioration of its accuracy. The purpose of this paper is to study the behavior of the structure under such effects and to calculate the changes in accuracy caused by them.

2. Nuclear magnetic gyroscope

The NMG is a quantum gyroscope the sensitive element of which is an ensemble of oriented atomic nuclei with a macroscopic magnetic moment.

2.1. NMG operation principles

The principle of NMG operation is as follows: nuclear spins are polarized under the influence of a magnetic field and / or resonant polarized light (pumping light); nuclear spins precess freely for some time; information about the orientation of the spins is read out by resonant (detecting) light. Usually, these processes take place at the same time and continuously, which allows continuous operation of the device. The dependence of NMG readings on the magnetic field can be largely eliminated by using two nuclear isotopes with different gyromagnetic ratios in one cell [1].

Our NMG uses spin exchange pumping. Spin exchange pumping can be carried out in a system containing a mixture of alkali metal and noble gas atoms. The working cell of such a gyroscope contains one or more NMR isotopes (for example, 129Xe and 131Xe), alkali metal atoms (Cs or 87Rb, hereinafter Rb), and buffer gas. Rubidium atoms are pumped by circularly polarized light. Spin polarization is transferred from cesium to xenon through the mechanism of spin exchange in collisions.

The spin precession for each of the two types of xenon atoms is synchronized by resonant alternating fields perpendicular to the direction of the constant field. The Cs precession frequency is about a thousand times higher than that of the Xe. Cs precession does not have any effect on the Xe atoms, while the Xe nuclear spin precession creates a slowly rotating field of B_{Xe}, which affects the electron spin of Cs atoms. Thus, information on the Xe nuclei precession can be obtained from the information on Cs magnetization; the latter is detected by a change in the polarization of the detection beam. An additional advantage of this method is the aforementioned effect of strengthening the effective magnetic field created by nuclear spins [2].

2.2. Basics of the structure

The basic principle of the NMG structure can be observed in Fig. 1. A quartz cell (1) with an ensemble of atoms is located in the center of the structure. This cell must be in a constant magnetic field with minimal disturbances, therefore, it is surrounded by a magnetic shield (2) and coils (3). Optical pumping is performed using a laser beam with a source (4). The data is read out using a detection beam (5) and a photodetector (6).

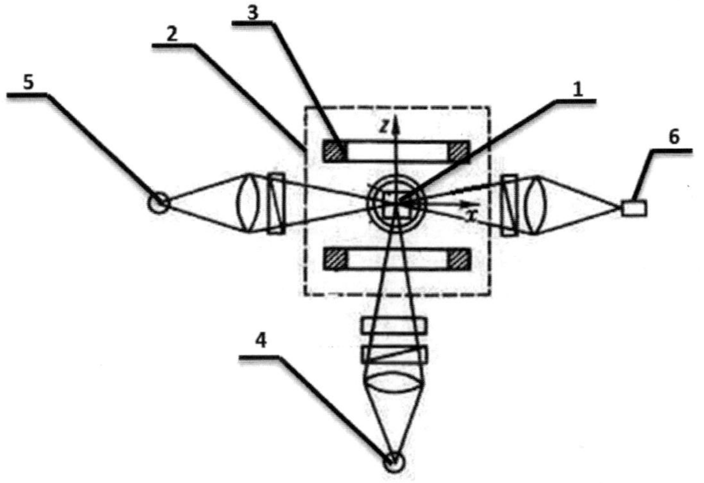

Figure 1. Schematic diagram of NMG.

The schematics of the NMG operation principle and the gyroscope signal are shown in Fig. 2.Information about the increment of the sensor rotation angle can be extracted from the phase of the Xe precession signal. In the case that two Xe isotopes are used in a cell, it is obtained from the relative phase of their precession. In this latter case, the sum of the precession frequencies of the Xe isotopes is used as an input signal for the magnetic field stabilization circuit.

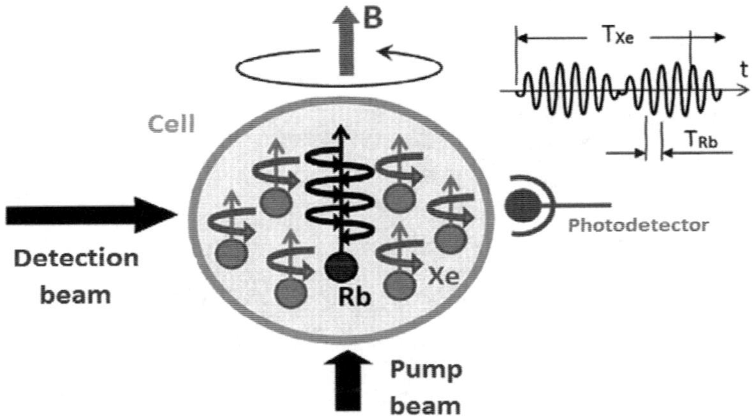

Figure 2. Principle of operation and schematic of the gyroscope signal.

2.3. NMG features
In recent years, interest in NMG has been reviving, which is connected, on the one hand, with the demand for miniature navigation-grade gyroscopes for small-sized navigation systems of highly dynamic objects that cannot be satisfied with the existing types of gyroscopes. On the other hand, it is connected with the advent of new technologies that offer prospects for NMG miniaturization [3, 4].

Table 1 shows the characteristics of modern types of gyroscopes that are most significant for consumers as well as the approximate requirements for different applications [5].

Table 1. Comparative characteristics of gyroscopes

Gyroscope type	Volume, cm^3	Zero offset instability, °/h	Range, °/s	Vibration resistance
Micromechanical	5	0.5	± 400	+
Fiber optic	600	0.001	± 40	+
Ring laser	250	0.001	± 40	++
Nuclear magnetic gyroscope	10	No worse than 0.01	± 400	+++

The characteristics indicated in the table are close to the maximum achievable level determined by the fundamental and technical limitations of existing technologies. As can be seen from the table, none of the currently existing types of gyroscopes has the required set of qualities. Thus, the main advantage of NMG over other types of gyroscopes is its fundamental insensitivity to mechanical effects.

3. NMG drift

3.1. Cause of drift
Consider the main causes leading to the NMG drift. From the aforesaid, we can conclude that the gyroscope readings are stable if the parameters of the pumping and detection radiation, as well as the magnetic field inside the case, do not change [6, 7, 8]. Thus, NMG drift depends on variations of the listed parameters relative to their nominal values.

- Working cell temperature (T);
- Pump-radiation parameters
  - Pump laser power (P_{pump});
  - Pump laser frequency (f_{pump});
- Detecting-radiation parameters
  - Detection laser power (P_{det});
  - Detection laser frequency (f_{det});
- Cell shape asymmetry (C);
- Magnetic field in the cell (B);

Bearing all this in mind, we can derive a formula that connects zero offsets with the parameters described above [2].

$$\sigma_{DRIFT} = 7 * 10^{-3}{}^{\circ}/h/mK * \Delta T + 2.5 * 10^{-5}{}^{\circ}/h/ppm * \Delta P_{pump} + 0.5^{\circ}/h/MHz * \Delta f_{pump} + 6 * 10^{-3}{}^{\circ}/h/ppm * \Delta P_{det} + 3^{\circ}/h/MHz * \Delta f_{det} + 1.5 * 10^{-3}{}^{\circ}/h/ppm * \Delta C + 2^{\circ}/h/pT * B \quad (1)$$

Based on this formula, Table 2 shows the variations of the parameters, that should not be exceeded in order to achieve stability at a level of 10^{-2} °/h [2].

Table 2. Variations of the parameters

Parameter	**Designation**	**Variations**
Working cell temperature	ΔT	1.4 mK
Pump laser power	ΔP_{pump}	400 ppm
Pump laser frequency	Δf_{pump}	0.02 MHz
Detection laser power	ΔP_{det}	1.7 ppm
Detection laser frequency	Δf_{det}	0.003 MHz
Cell shape asymmetry	ΔC	6.7 ppm
Magnetic field in the cell	ΔB	5 pT

Under various physical loads, the pump and detection lasers may shift. Thus, the volume of atoms to which the energy of the lasers is imparted will change and the power will also change. Since the other laser parameters do not change in this case, the ratio of laser powers can be considered as the ratio of the volume formed by the intersection of two beams. Initially, this volume is equal to one mm^3. The temperature of the cell, as well as the frequency of laser radiation, does not change in the course of mechanical effects. Meanwhile, the changes in the cell shapes due to the liquid metal evaporation and condensation inside the cell are much more noticeable than those caused by deformations due to external effects.

3.2. Design model

A finite element model was built in ProEngineer to obtain information on displacements. Displacements occur in the structure in the presence of physical effects. The NMG model developed by the designers of Concern CSRI Elektropribor was taken as a basis.

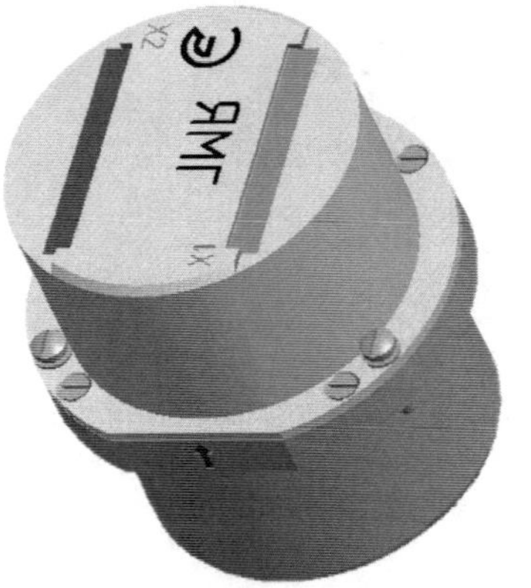

Figure 3. Model of NMG.

To create boundary conditions and to further estimate the stress-strain state in the specialized CAD system ProEngineer, a geometric computer model was built, which was further used in the finite element analysis (FEA) program in the modules for dividing the model into finite elements. The geometric model contains all the dimensions of the structural elements and follows their shapes.

For the whole object, even with its complex geometry, the area may be split into elements during automatic meshing. In this work, tetrahedrons were used as such elements. Owing to a smart selection algorithm, you can create a mesh that best matches the real geometry of the part.

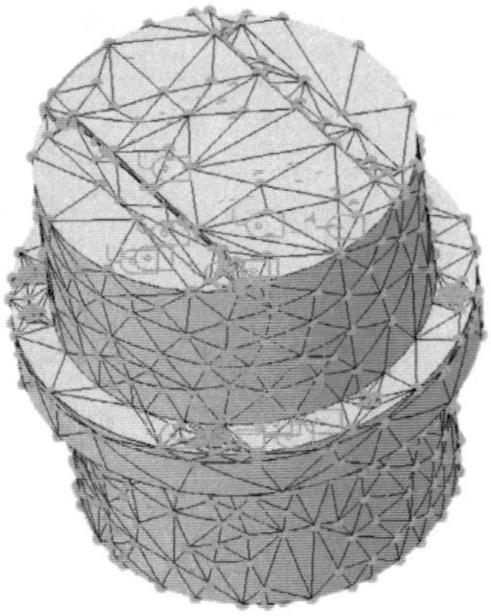

Figure 4. Model's mesh.

As a result of the mesh construction, the number of volumetric elements in the model was 28 909 pieces.

3.3. Calculation based on the constructed model

To calculate the response of the model to various effects, first of all, it is necessary to carry out a modal analysis to detect the proper frequencies of the structure.

The range of the proper frequency search is 0–3000Hz, based on the technical requirements of the device. As a result, the resonant frequencies were 1229, 2265, 2519, 2857, and 2978 Hz. The next step was calculation of mechanical effects. The vibrations were selected with an amplitude of 20g in the range from 0 to 3000 Hz. This impact is directed along the X- and Y-axes indicated in Fig. 5.

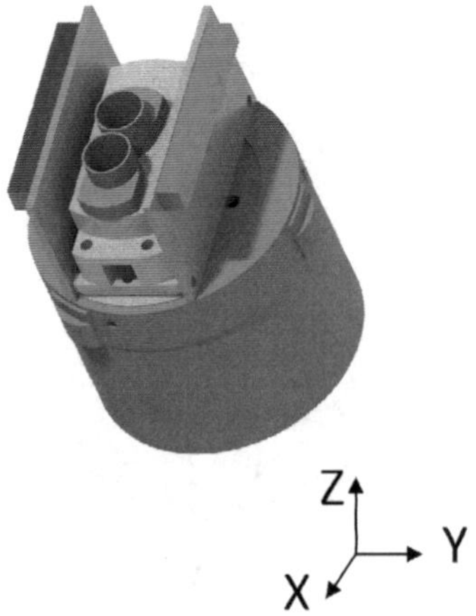

Figure 5. Direction of the axes in the simulation.

As a result, the following graphs were obtained for the offsets of the pump and detection lasers, including the cell, along the axis of exposure:

IOP Conf. Series: Materials Science and Engineering 1215 (2022) 012014 doi:10.1088/1757-899X/1215/1/012014

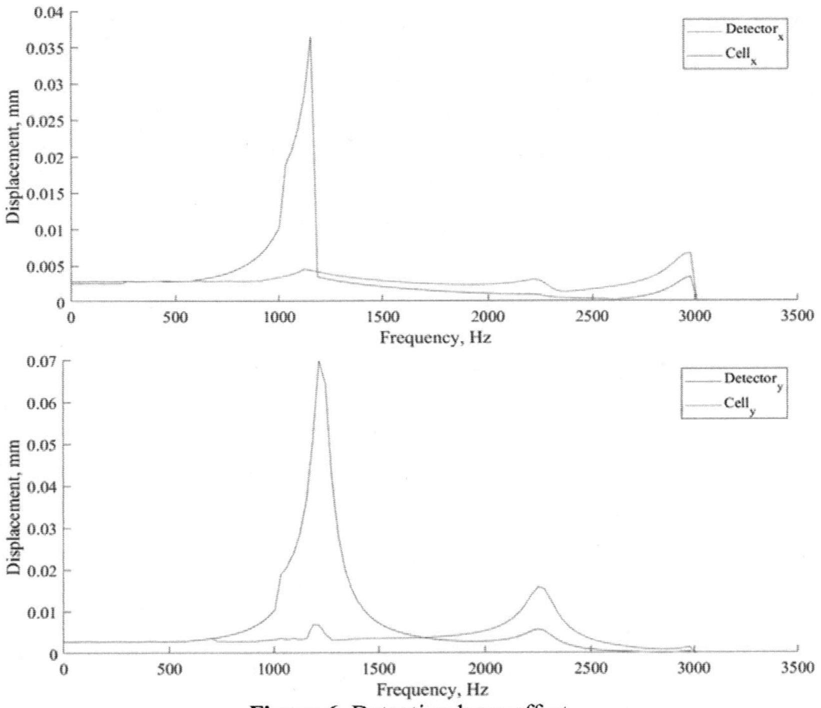

Figure 6. Detection laser offset.

It should be noted that a damping factor of 10% is taken in the calculations. This coefficient requires clarification based on the results of a full-scale test of the device. If we change the damping factor, there will also be a change in the height of the resonance peaks.

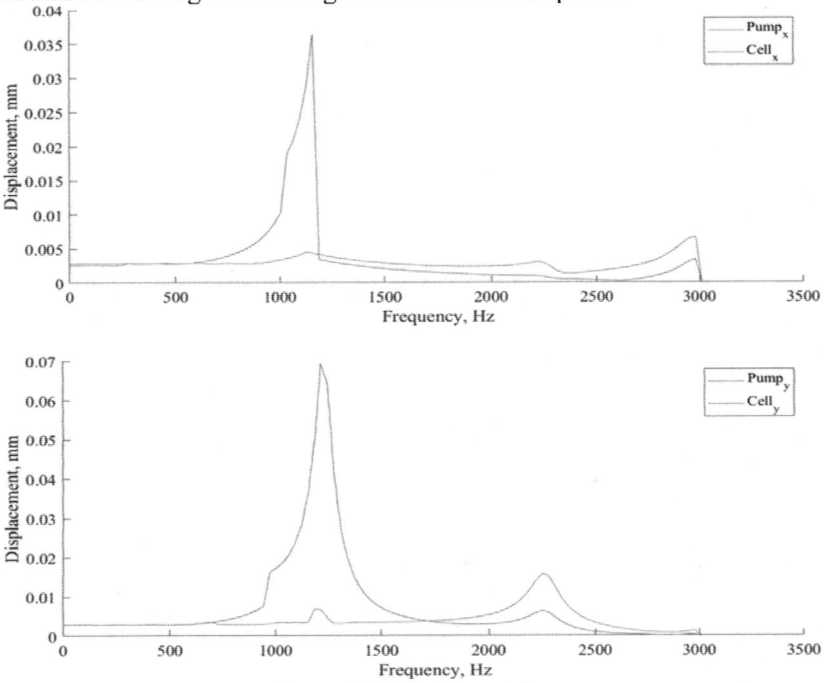

Figure 7. Pump laser offset.

Based on the data of laser displacements, the change in the volume of oriented atoms is calculated. In the calculations, it was taken into account that the laser diameter is 1 mm, i.e., initially the volume of oriented atoms was approximately 1 mm^3. By the changes in the volume, one can also judge about the changes in the laser power supplied to the new volume. The ratio of these powers is the same for the pump and detection lasers. This value can be directly supplied to Equation 1. As a consequence, the following value of the zero offset caused by vibration effects on the device was obtained.

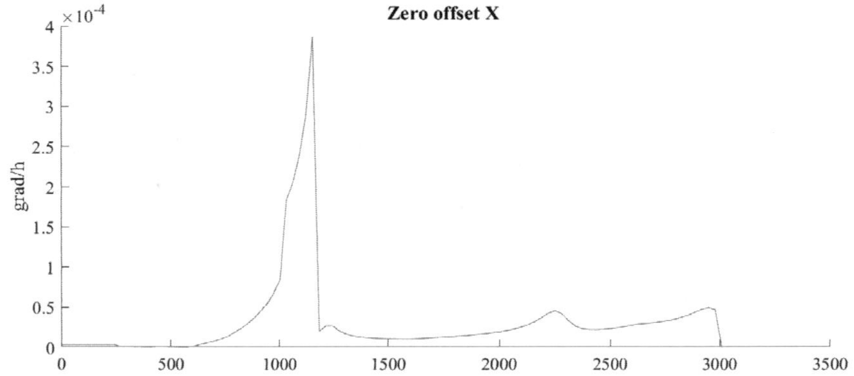

Figure 8. Drift under the impact along the X-axis.

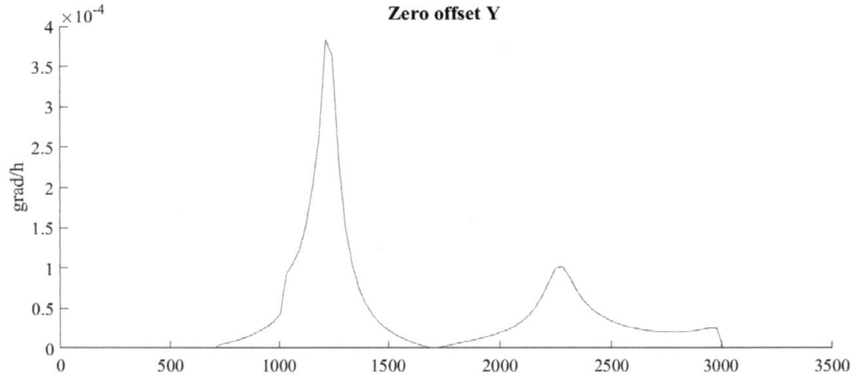

Figure 9. Drift under the impact along the Y-axis.

Based on the above graphs, Table 3 was compiled, showing the values of the maximum zero offsets for the resonant frequencies.

Table 3. Zero offset of NMG

Resonant frequency	**Maximum zero offset value, 10^{-4} °/h**	
	X	**Y**
1229	3.9	3.6
2265	0.4	0.9
2519	0.2	0.3
2857	0.4	0.2
2978	0.5	0.2

From the table, we can conclude that the deformations arising from inertial effects cause errors in the gyroscope operation of the order of 10^{-3} °/h.

4. Conclusions

- A finite element model of NMG has been constructed.
- The simulation of the inertial effects has been carried out.
- The gyroscope zero offset based on the offsets caused by inertial forces have been calculated.

As a result, the following conclusions can be drawn:

- The NMG structure revealed the presence of resonances in the frequency range of up to 3 kHz.
- Zero offsets caused by the offsets at resonant frequencies are less than 10^{-3} °/h.
- The error may increase with an increase in the damping coefficient values.

References

[1] Feng D. Review of Quantum navigation. IOP Conference Series: Earth and Environmental Science. 2019; 237:032027

[2] Vershovskii A, Litmanovich Y, Pazgalev A, Peshekhonov V. Nuclear Magnetic Resonance Gyro: Ultimate Parameters. *Gyroscopy Navig.* **9** 162-176 (2018).

[3] Fang J, Qin J. Advances in Atomic Gyroscopes: A View from Inertial Navigation Applications. *Sensors* **12** 6331-6346 (2012).

[4] Liu Y, Shi M, Wang X. Progress on atomic gyroscope. 24th Saint Petersburg International Conference on Integrated Navigation Systems (ICINS). 2017.

[5] Peshekhonov, V.G. The Outlook for Gyroscopy. *Gyroscopy Navig.* **11**, 193–197 (2020). https://doi.org/10.1134/S2075108720030062

[6] Simpson J, Fraser J, Greenwood I. An Optically Pumped Nuclear Magnetic Resonance Gyroscope. IEEE Transactions on Aerospace. 1963 ;1(2):1107-1110.

[7] Walker T, Larsen M. Spin-Exchange-Pumped NMR Gyros. Advances In Atomic, Molecular, and Optical Physics. 2016; 373-401.

[8] Larsen M, Bulatowicz M. Nuclear Magnetic Resonance Gyroscope: For DARPA's micro-technology for positioning, navigation and timing program. 2012 IEEE International Frequency Control Symposium Proceedings. 2012.

AUTHOR INDEX